T0269512

CAMBRIDGE LIBRARY COLLECTION

Books of enduring scholarly value

Botany and Horticulture

Until the nineteenth century, the investigation of natural phenomena, plants and animals was considered either the preserve of elite scholars or a pastime for the leisured upper classes. As increasing academic rigour and systematisation was brought to the study of 'natural history', its subdisciplines were adopted into university curricula, and learned societies (such as the Royal Horticultural Society, founded in 1804) were established to support research in these areas. A related development was strong enthusiasm for exotic garden plants, which resulted in plant collecting expeditions to every corner of the globe, sometimes with tragic consequences. This series includes accounts of some of those expeditions, detailed reference works on the flora of different regions, and practical advice for amateur and professional gardeners.

Flora Domestica

Elizabeth Kent (1790–1861) lived in London, but wanted to live in the country. Dismayed at the number of pot-plants given to her which failed to thrive, she published this useful guide to container or 'portable' gardening in 1823. She had taught herself botany and foreign languages, and her sister's marriage to the radical poet and journalist Leigh Hunt brought her into contact with the Romantic circles. The book combines practical instruction on how to select plants which will thrive in containers, and in the polluted air of cities, with quotations on gardening and flowers from ancient as well as modern authors such as Keats and her friend Shelley. Her common-sense advice on plants from adonis to zygophyllum and on their care – use rainwater if possible, but never overwater or let pots stand in water, for example – is equally valid today.

Cambridge University Press has long been a pioneer in the reissuing of out-of-print titles from its own backlist, producing digital reprints of books that are still sought after by scholars and students but could not be reprinted economically using traditional technology. The Cambridge Library Collection extends this activity to a wider range of books which are still of importance to researchers and professionals, either for the source material they contain, or as landmarks in the history of their academic discipline.

Drawing from the world-renowned collections in the Cambridge University Library and other partner libraries, and guided by the advice of experts in each subject area, Cambridge University Press is using state-of-the-art scanning machines in its own Printing House to capture the content of each book selected for inclusion. The files are processed to give a consistently clear, crisp image, and the books finished to the high quality standard for which the Press is recognised around the world. The latest print-on-demand technology ensures that the books will remain available indefinitely, and that orders for single or multiple copies can quickly be supplied.

The Cambridge Library Collection brings back to life books of enduring scholarly value (including out-of-copyright works originally issued by other publishers) across a wide range of disciplines in the humanities and social sciences and in science and technology.

Flora Domestica

Or the Portable Flower-Garden

ELIZABETH KENT

CAMBRIDGE
UNIVERSITY PRESS

CAMBRIDGE
UNIVERSITY PRESS

University Printing House, Cambridge, CB2 8BS, United Kingdom

Cambridge University Press is part of the University of Cambridge.
It furthers the University's mission by disseminating knowledge in the pursuit of education, learning and research at the highest international levels of excellence.

www.cambridge.org
Information on this title: www.cambridge.org/9781108076739

This edition first published 1823
This digitally printed version 2017

ISBN 978-1-108-07673-9 Paperback

FLORA DOMESTICA.

LONDON:
PRINTED BY THOMAS DAVISON, WHITEFRIARS.

FLORA DOMESTICA,

OR

THE PORTABLE FLOWER-GARDEN;

WITH

DIRECTIONS FOR THE TREATMENT OF

PLANTS IN POTS;

AND

ILLUSTRATIONS FROM THE WORKS OF THE POETS.

"How exquisitely sweet
This rich display of flowers,
This airy wild of fragrance
So lovely to the eye,
And to the sense so sweet."

ANDREINI'S ADAM.

LONDON:
PRINTED FOR TAYLOR AND HESSEY,
93, FLEET-STREET,
AND 13, WATERLOO-PLACE, PALL-MALL.

1823.

TO

SIR WILLIAM KNIGHTON, BART.

SIR,

I TAKE the liberty of laying this Volume before you, in humble acknowledgment of the gratitude and respect with which I remain,

SIR,

Your humble and obedient Servant,

THE AUTHOR.

A

LIST OF THE PLANTS

DESCRIBED IN THIS WORK.

PREFACE.

As I reside in town, and am known among my friends as a lover of the country, it has often happened that one or other of them would bring me consolation in the shape of a Myrtle, a Geranium, an Hydrangea, or a Rose-tree, &c. Liking plants, and loving my friends, I have earnestly desired to preserve these kind gifts; but, utterly ignorant of their wants and habits, I have seen my plants die one after the other, rather from attention ill-directed than from the want of it. I have many times seen others in the same situation as myself, and found it a common thing, upon the arrival of a new plant, to hear its owner say, "Now, I should like to know how I am to treat this? Should it stand within doors, or without? should it have much water, or little? should it stand in the sun, or in the shade?"

Even Myrtles and Geraniums, commonly as they are seen in flower-stands, balconies, &c., often meet with an untimely death from the ignorance of their nurses. Many a plant have I destroyed, like a fond and mistaken mother, by an inexperienced tenderness; until, in pity to these vegetable nurslings and their nurses, I resolved to obtain and to communicate such information as should be requisite for the rearing and preserving a *portable garden* in pots. This little volume is the result; the information contained in it has been carefully collected from the best authorities; and henceforward the death of any plant, owing to the carelessness or ignorance of its nurse, shall be brought in, at the best, as *plant-slaughter.*

It has not been attempted to make a complete catalogue of every plant that may be reared in a pot or tub, but such have been selected as are the most frequently so cultivated; and such as are most desirable for beauty of form or colour, luxuriance of foliage, sweetness of perfume, or from interesting or poetical associations with their history. In the belief that lovers of nature are most frequently admirers of beauty in any form, such anecdotes or poetical passages are added, relating to the plants mentioned, as appeared likely to interest them.

To avoid endless repetition, some few general observations are subjoined, but only such as are really general; and they will not be found to render a variety of references necessary for the treatment of one plant, a necessity which it is the chief aim of this little work to set aside. It is hoped that any person desiring to know the treatment proper for this or that plant, will find all the information necessary under its particular head. The *General Observations* are comprised in so small a compass, that the merely reading them over will probably be found sufficient.

The love of flowers is a sentiment common alike to the great and to the little; to the old and to the young; to the learned and the ignorant, the illustrious and the obscure. While the simplest child may take delight in them, they may also prove a recreation to the most profound philosopher. Lord Bacon himself did not disdain to bend his mighty intellect to the subject of their culture.

Lord Burleigh also found recreation from the cares of state in his flower-garden. Ariosto, although utterly ignorant of botanical science, took even an infantine pleasure in his little garden; and we are informed by his son, that after sowing a variety of seeds, he would watch eagerly for the springing of the plants, would cherish the first peep of vegetation, and having for many days watered and tended the young plant, discover at last that he had bestowed all this tenderness upon a weed; a weed, perhaps, which had choked the plant for which he had mistaken it.

"Nelle cose de' giardini teneva il modo medesimo, che nel far de versi, perche mai non lasciava cosa alcuna che piantasse più di tre mesi in un loco'; e se piantava anime di persiche, o semente di alcuna sorte, andava tante volte a vedere se germogliava, che finalmente, rompea il germoglio: e perche avea poco cognizione d' erba, il più delle volte prossumea che qualunque erba, che nascesse vicina alla cosa seminata da esso, fosse quella; la custodiva con diligenza grande sin tanto che la cosa fosse ridotta a' termini, che, non accascava averne dubbio. I' mi ricordo, ch' avendo seminato de' capperi, ogni giorno andava a vederli, e stava con una allegrezza grande di cosi bella nascione. Finalmente trovo ch' eran sambuchi, e che de' capperi non n'eran nati alcuni."

"He treated his garden as he did his verses, never leaving any thing three months in the same place. Whenever he planted or sowed any thing, he went so often to see if it sprouted, that at last he broke the shoot: and having little knowledge of plants, he took any leaves that appeared near the place where he had sown his seeds for the plants sown, and tended them with the greatest diligence, till his mistake was clear beyond doubt. I remember once when he had sown some capers, he went every day to look at them, and was delighted to see them thrive so well. At last he found these thriving plants were young elders, and that none of the capers had appeared."

NOTES BY VIRGINIO ARIOSTO, FOR A LIFE OF HIS FATHER.

Who can read this anecdote of so great a man, and not feel an additional interest in him! In how amiable a light it represents him! Was a cruel, unfeeling, or selfish man ever known to take pleasure in working in his own garden? Surely not. This love of nature in detail (if the expression may be allowed) is an union of affection, good taste, and natural piety.

How amiable a man was Cowper!—and Evelyn, too, and Evelyn's friend, Cowley, who addressed to him a poem entitled The Garden. Gessner also is represented as of a kindred sweetness of nature. They all worked in their own gardens; and with enthusiastic pleasure.

Barclay, the author of the Argenis, rented a house near the Vatican, in Rome, with a garden in which he planted the choicest flowers, principally such as grow from bulbs, which had never been seen in Rome before. He was extremely fond of flowers, particularly of the bulbous kind, which are prized chiefly for their colours, and purchased the bulbs at a high price*.

Pope had the same taste, and was assisted in his horticultural amusements by Lord Peterborough. One of the most interesting descriptions of him represents him as being seen before dinner in a small suit of black, very neat and gentlemanly, with a basket in his hand containing flowers for the Miss Blounts. Rousseau, who has written some interesting Letters on Botany, of which among his other accomplishments he was master, found friends in the flowers, when he thought he had no others. Even his great rival Voltaire, who if he had more wit had much less sentiment, soothed his irritability and cherished his benevolence in his garden; and one, " greater than he," and whom I mention in the same page with any thing but an irreverent or unchristian feeling, said the noblest thing of a flower that ever was uttered: "Behold the lilies of the field, how they grow: they toil not, neither do they spin; and yet I say unto you, that Solomon, in all his glory, was not arrayed like one of these." (Matthew, chap. vi. v. 28, 29†.) How surely would Solomon himself have

* See Beckmann's History of Inventions, vol. i.

† Some have supposed that the flower to which Jesus alluded must have been the Tulip; as if it were necessary for it to be really gaudy or gorgeous before it could be set above the splendour of royalty! This may be called the art of divesting sentiment of its sentiment.

agreed with this beautiful speech! for that his "wise heart" loved the flowers, the lily especially, is evident from numerous passages in his Song. The object of his love in claiming a supreme degree of beauty, exclaims, "I am the rose of Sharon, and the lily of the valley."

The Emperor Dioclesian preferred his garden to a throne:

"Methinks I see great Dioclesian walk
In the Salonian garden's noble shade,
Which by his own imperial hands was made:
I see him smile, methinks, as he does talk
With the ambassadors, who come in vain
T' entice him to a throne again.
'If I, my friends,' said he, 'should to you show
All the delights which in these gardens grow,
'Tis likelier far that you with me should stay,
Than 'tis that you should carry me away:
And trust me not, my friends, if, every day,
I walk not here with more delight,
Than ever, after the most happy fight,
In triumph to the capital I rode,
To thank the gods, and to be thought myself almost a god.'"

COWLEY'S GARDEN.

Sir W. Temple desired to have his heart buried in his garden.

Lope de Vega appears to have been a lover of gardens. "As he is mentioned more than once," says Lord Holland, "by himself and his encomiasts, employed in trimming a garden, we may collect that he was fond of that occupation. Indeed his frequent description of parterres and fountains, and his continual allusion to flowers, justify his assertion—'that his garden furnished him with ideas, as well as vegetables and amusement*.'"

The French poet Ronsard was evidently a lover of

* See Life of Lope de Vega, vol. i. page 93.

flowers, as may be seen in his poems, particularly of the Rose, and the Violet, which he calls the flower of March; these he has introduced repeatedly:

"Two flowers I love, the March-flower and the rose,
The lovely rose that is to Venus dear *."

Ovid was, as might be expected, a lover of gardens, and by a passage in one of his poems appears to have been fond of writing in them. It is in his Tristia, where he is regretting, during his voyage to the place of his exile, the delight he used to feel in composing his verses under the genial sky, and among the domestic comforts of his native country:

"Non hæc in nostris, ut quondam, scribimus hortis,
Nec, consuete, meum, lectule, corpus habes:
Jactor in indomito brumali luce profundo,
Ipsaque cæruleis charta feritur aquis.
Improba pugnat hiems, indignaturque, quòd ausim
Scribere, se rigidas incutiente minas."

Lib. i. Eleg. 11.

"Not in my garden, as of old, I write,
With thee, dear couch, to finish the delight:
I toss upon a ghastly wintery sea,
While the blue sprinkles dash my poetry.
Fell winter's at his war; and storms the more
To see me dare to write for all his threatening roar."

Ovid is so fond of flowers, that, in the account of the Rape of Proserpine in his Fasti, he devotes several lines to the enumeration of the flowers gathered by her attendants. Mr. Gibbon is very angry with him for it: "Can it be believed," says he, "that the Rape of Proserpine should be described in two verses, when the enumeration of the flowers which she gathered in the garden of Eden had just

* See Mr. Cary's Translation in the London Magazine, vol. v. page 507.

filled sixteen*?" But surely this loitering of the poet, over his meadows and crocuses, conveys a fit sense of the pleasure enjoyed by Proserpine and her nymphs; a pleasure, too, for which they expressly came forth, and by the too great pursuit of which the latter were separated from their mistress.

In our own time, we may instance the late Mr. Shelley. Of a strong and powerful intellect, his manners were gentle as a summer's evening: his tastes were pure and simple: it was his delight to ramble out into the fields and woods, where he would take his book, or sometimes his pen, and having employed some hours in study, and in speculations on his favourite theme—the advancement of human happiness, would return home with his hat wreathed with briony, or wild convolvulus; his hand filled with bunches of wild-flowers plucked from the hedges as he passed, and his eyes, indeed every feature, beaming with the benevolence of his heart. He loved to stroll in his garden, chatting with a friend, or accompanied by his Homer or his Bible (of both which he was a frequent reader): but one of his chief enjoyments was in sailing, rowing, or floating in his little boat, upon the river: often he would lie down flat in the boat and read, with his face upwards to the sunshine. In this taste for the water he was too venturesome, or perhaps inconsiderate; for it was rather a thoughtlessness of danger, than a braving of it. In the end, as it is well known, it was fatal to him: never will his friends cease to feel, or to mourn his loss; though their mourning will be softened by the contemplation of his amiable nature, and by the memory of that gentle and spiritual countenance, "which seemed not like an inhabitant of the earth" while it was on it.

* Gibbon's Miscellaneous Works, vol. iv. page 356.

Among the existing lovers of flowers, it is a pleasure to be able to name the gallant and accomplished young prince, Alexander Mavrocordato, one of the chief leaders of the Greeks in their present glorious struggle for freedom. A botanical work, not long since published in Italy, is dedicated to him on account of his known fondness for the subject. Thus, in every respect, he inherits the feelings of his ancestors. This is the same prince to whom Mr. Shelley dedicated his Hellas. Among the Greeks this taste was very general, as may be gathered from many ancient writers. In the following passage from the Travels of Anacharsis, several of these authorities are assembled: the author describes a visit to a friend who had retired to his country-house:

"Après avoir traversé une basse-cour peuplée de poules, de canards, et d'autres oiseaux domestiques, nous visitâmes l'écurie, la bergerie, ainsi que le jardin des fleurs, où nous vimes successivement briller les narcisses, les jacinthes, les anemones, les iris, les violettes de différentes couleurs, les roses de diverses espèces, et toutes sortes de plantes odoriférantes. Vous ne serez pas surpris, me dit-il, du soin que je prends de les cultiver: vous savez que nous en parons les temples, les autels, les statues de nos dieux; que nous en couronnons nos têtes dans nos repas et dans nos ceremonies saintes; que nous les repandons sur nos tables et sur nos lits; que nous avons même l'attention d'offrir à nos divinités les fleurs qui leur sont les plus agréables. D'ailleurs un agriculteur ne doit point négliger les petits profits; toutes les fois que j'envoie au marché d'Athènes, du bois, du charbon, des denrées et des fruits, j'y joins quelques corbeilles de fleurs qui sont enlevées à l'instant *."

"Having crossed a court-yard peopled with fowls, ducks, and other domestic birds, we visited the stables, the sheep-fold, and the flower-garden; where we saw in succession narcissuses, hyacinths, anemonies, irises, violets of different colours, roses of various kinds, and all sorts of odoriferous plants. You will not be surprised, said he, at the care I take in cultivating them; for you know that we adorn with them

* Voyage du Jeune Anacharsis en Grèce, vers le milieu du quatrième siècle avant l'ere vulgaire; par J. J. Barthélemy. Tome cinquième.

the temples, altars, and statues of our gods; that we crown our heads with them in our festivals, and holy ceremonies; that we scatter them upon our tables, and our beds; that we even consider the kinds of flowers most agreeable to our divinities. Besides an agriculturist should not neglect small profits; whenever I send to the market of Athens wood, provision, or fruit, I add some baskets of flowers, and they are seized instantly."

In another part of the same work, the author describes a marriage ceremony in the Island of Delos, in which flowers, shrubs, and trees make a conspicuous figure. He tells us that the inhabitants of the island assembled at day-break, crowned with flowers: that flowers were strewed in the path of the bride and bridegroom: the house was garlanded with them: singers and dancers appeared, crowned with oak, myrtle, and hawthorns; the bride and bridegroom were crowned with poppies; and upon their approach to the temple a priest received them at the entrance, presenting to each a branch of ivy,—a symbol of the tie which was to unite them for ever *.

It was not in their sports only that the Greeks were so lavish of their flowers: they crowned the dead with them; and the mourners wore them in the funeral ceremonies. Flowers seem to have been to this tasteful people a sort of poetic language, whereby they expressed the intensity of feelings to which they found common language inadequate. Thus we find that their grief, and their joy, their religion, and their sports, their gratitude, admiration, and love, were alike expressed by flowers.

And flowers do speak a language, a clear and intelligible language: ask Mr. Wordsworth, for to him they have spoken, until they excited "thoughts that lie too deep for tears;" ask Chaucer, for he held companionship with them in the meadows; ask any of the poets, ancient or modern. Observe them, reader, love them, linger over

* Vol. vi. chapter 77.

them; and ask your own heart if they do not speak affection, benevolence, and piety. None have better understood the language of flowers than the simple-minded peasant-poet, Clare, whose volumes are like a beautiful country, diversified with woods, meadows, heaths, and flower-gardens: the following is a pleasing specimen:

"Bowing adorers of the gale,
Ye cowslips delicately pale,
Upraise your loaded stems;
Unfold your cups in splendour, speak!
Who decked you with that ruddy streak,
And gilt your golden gems?

"Violets, sweet tenants of the shade,
In purple's richest pride arrayed,
Your errand here fulfil;
Go bid the artist's simple stain
Your lustre imitate, in vain,
And match your Maker's skill.

"Daisies, ye flowers of lowly birth,
Embroiderers of the carpet earth,
That stud the velvet sod;
Open to spring's refreshing air,
In sweetest smiling bloom declare
Your Maker, and my God *."

This poet is truly a lover of Nature: in her humblest attire she still is pleasing to him, and the sight of a simple weed seems to him a source of delight:

"There 's many a seeming weed proves sweet,
As sweet as garden-flowers can be †."

In his lines to Cowper Green, he celebrates plants that seldom find a bard to sing them; having enumerated several, he continues;—

"Still thou ought'st to have thy meed,
To show thy flower as well as weed.

* Clare's Village Minstrel and other Poems, vol. ii. page 61.
† Clare's Poems on Rural Life, &c. page 63.

Though no fays, from May-day's lap,
Cowslips on thee dare to drop;
Still does nature yearly bring
Fairest heralds of the spring:
On thy wood's warm sunny side
Primrose blooms in all its pride;
Violets carpet all thy bowers;
And anemone's weeping flowers,
Dyed in winter's snow and rime,
Constant to their early time,
White the leaf-strewn ground again,
And make each wood a garden then.
Thine's full many a pleasing bloom
Of blossoms lost to all perfume:
Thine the dandelion flowers,
Gilt with dew, like suns with showers;
Harebells thine, and bugles blue,
And cuckoo flowers all sweet to view;
Thy wild-woad on each road we see;
And medicinal betony,
By thy woodside railing, reeves
With antique mullein's flannel leaves.
These, though mean, the flowers of waste,
Planted here in nature's haste,
Display to the discerning eye
Her loved, wild variety:
Each has charms in nature's book
I cannot pass without a look.
And thou hast fragrant herbs and seed,
Which only garden's culture need:
Thy horehound tufts, I love them well,
And ploughman's spikenard's spicy smell;
Thy thyme, strong-scented 'neath one's feet;
Thy marjoram beds, so doubly sweet;
And pennyroyals creeping twine:
These, each succeeding each, are thine,
Spreading o'er thee wild and gay,
Blessing spring, or summer's day.
As herb, flower, weed, adorn thy scene,
Pleased I seek thee, Cowper Green."

VILLAGE MINSTREL, &c. vol. i. page 113.

The eloquence of flowers is not perhaps so generally

understood in this country as it might be, but Mr. Bowring scarcely does us justice in the following observations:

"In the peninsula the wildest flowers are the sweetest. There are hedges of myrtles, and geraniums, and pomegranates, and towering aloes. The sunflower and the bloody warrior (Aleli grosero) occupy the parterre: they are no favorites of mine.

"Flowers! what a hundred associations the word brings to my mind. Of what countless songs, sweet and sacred, delicate and divine, are they the subject. A flower in England is something to the botanist,—but only if it be rare; to the florist—but only if it be beautiful; even the poet and the moralizer seldom bend down to its eloquent silence. The peasant never utters to it an ejaculation—the ploughman (all but one) carelessly tears it up with his share—no maiden thinks of wreathing it—no youth aspires to wear it. But in Spain ten to one but it becomes a minister of love, that it hears the voice of poetry, that it crowns the brow of beauty. Thus how sweetly an anonymous cancionero sings:

"Put on your brightest, richest dress,
Wear all your gems, blest vales of ours!
My fair one comes in her loveliness,
She comes to gather flowers.

"Garland me wreaths, thou fertile vale;
Woods of green your coronets bring;
Pinks of red, and lilies pale,
Come with your fragrant offering.
Mingle your charms of hue and smell,
Which Flora wakes in her spring-tide hours!
My fair one comes across the dell,
She comes to gather flowers.

"Twilight of morn! from thy misty tower
Scatter the trembling pearls around,

Hang up thy gems on fruit and flower,
Bespangle the dewy ground!
Phœbus, rest on thy ruby wheels—
Look, and envy this world of ours;
For my fair one now descends the hills,
She comes to gather flowers.

"List! for the breeze on wings serene
Through the light foliage sails;
Hidden amidst the forest green
Warble the nightingales!
Hailing the glorious birth of day
With music's best, divinest powers,
Hither my fair one bends her way,
She comes to gather flowers."

London Magazine, *Spanish Romances,* No. 3.

For the most part of our countrymen, I fear they do not allow themselves leisure to admire or enjoy the beauties of nature; yet it cannot be said that they are utterly insensible to them; for with regard to flowers at least we may observe, that on Sundays every village beau, nay every straggling townsman who comes on that day within reach of a flower, has one in his button-hole.

It was, perhaps, the general power of sympathy upon the subject of plants, which caused them to be connected with some of the earliest events that history records. The mythologies of all nations are full of them; and in all times they have been associated with the soldiery, the government, and the arts. Thus the patriot was crowned with oak; the hero and the poet with bay; and beauty with the myrtle. Peace had her olive; Bacchus his ivy; and whole groves of oak-trees were thought to send out oracular voices in the winds. One of the most pleasing parts of state-splendor has been associated with flowers, as Shakspeare seems to have had in his mind when he wrote that beautiful line respecting the accomplished prince, Hamlet:

"The expectancy and rose of the fair state."

It was this that brought the gentle family of roses into such unnatural broils in the civil wars: and still the united countries of Great Britain have each a floral emblem: Scotland has its thistle, Ireland its shamrock, and England the rose. France, under the Bourbons, has the golden lily.

It was an annual custom with the Popes to send a golden rose perfumed to the Prince who happened to be most in their good graces.

Our different festivals have each their own peculiar plant, or plants, to be used in their celebration: at Easter the willow as a substitute for the palm; at Christmas, the holly and the mistletoe; on May-day every flower in bloom, but particularly the hawthorn or May-bush. In Persia they have a festival called the Feast of Roses, which lasts the whole time they are in bloom—(See Roses, page 321). Formerly it was the custom, and still is in some parts of the country, to scatter flowers on the celebration of a wedding, a christening, or even of a funeral (See Roses, page 315, and Rosemary, page 331).

It was formerly the custom also to carry garlands before the bier of a maiden, and to hang them, and scatter flowers over her grave:

———————"Her death was dreadful;
And, but that great command o'ersways the order,
She should in ground unsanctified have lodged
Till the last trumpet; for charitable prayers,
Shards, flints, and pebbles should be thrown on her,
Yet here she is allowed her virgin crants*,
Her maiden strewments, and the bringing home
Of bell and burial."

The Queen scattering flowers:

"Sweets to the sweet. Farewell!
I hoped thy bride-bed to have decked, sweet maid,
And not have strewed thy grave."

Hamlet, Act v. Scene 1.

* Crants is the German word for garlands.

In Tripoli, on the celebration of a wedding, the baskets of sweetmeats, &c., sent as wedding presents, are covered with flowers; and although it is well known that they frequently communicate the plague, the inhabitants will even prefer running the risk, when that dreadful disease is abroad, rather than lose the enjoyment they have in their love of flowers. When a woman in Tripoli dies, a large bouquet of fresh flowers, if they can be procured, if not, of artificial, is fastened at the head of her coffin. Upon the death of a Moorish lady of quality, every place is filled with fresh flowers and burning perfumes: at the head of the body is placed a large bouquet, of part artificial, and part natural, and richly ornamented with silver: and additions are continually made to it. The author who describes these customs also mentions a lady of high rank, who regularly attended the tomb of her daughter, who had been three years dead: she always kept it in repair, and, with the exception of the great mosquè, it was one of the grandest in Tripoli. From the time of the young lady's death, the tomb had always been supplied with the most expensive flowers, placed in beautiful vases; and, in addition to these, a great quantity of fresh Arabian Jessamines, threaded on thin slips of the Palm-leaf, were hung in festoons and tassels about this revered sepulchre. The mausoleum of the royal family, which is called the Turbar, is of the purest white marble, and is filled with an immense quantity of fresh flowers; most of the tombs being dressed with festoons of Arabian Jessamine and large bunches of variegated flowers, consisting of Orange, Myrtle, Red and White Roses, &c. They afford a perfume which those who are not habituated to such choice flowers can scarcely conceive. The tombs are mostly of white, a few inlaid with coloured marble. A manuscript Bible, which was presented by a Jew to the Synagogue, was adorned with flowers; and silver vases filled with flowers were placed upon the ark which contained the sacred MS *.

The ancients used wreaths of flowers in their entertainments, not only for pleasure, but also from a notion that their odour prevented the wine from intoxicating them: they used other perfumes on the same account. Beds of flowers are not merely fictitious (see ROSES, page 320). The Highlanders of Scotland commonly sleep on heath, which is said to make a delicious bed; and beds are, in Italy, often filled with the leaves of trees, instead of down or feathers. It is an old joke against the effeminate Sybarites, that one of them complaining he had not slept all night, and being asked the reason why, said that a rose-leaf had got folded under him.

In Naples, and in the Vale of Cachemere (I have been told also that it sometimes occurs in Chester), gardens are formed on the roofs of houses: "On a standing roof of wood is laid a covering of fine earth, which shelters the building from the great quantity of snow that falls in the winter season. This fence communicates an equal warmth in winter, as a refreshing coolness in summer, when the tops of the houses, which are planted with a variety of flowers, exhibit at a distance the spacious view of a beautifully chequered parterre." (FORSTER.) The famous hanging gardens of Babylon were on the enormous walls of that city.

A garden usually makes a part of every Paradise, even of Mahomet's, from which women are excluded,—women, whom gallantry has so associated with flowers, that we are told, in the Malay language, one word serves for both †. In Milton's Paradise, the occupation of Adam and Eve

* See Tully's Narrative of a Residence in Tripoli.

† See Lalla Rookh, page 303. Sixth edition.

was to tend the flowers, to prune the luxuriant branches, and support the roses, heavy with beauty (see Roses, page 323). Poets have taken pleasure in painting gardens in all the brilliancy of imagination. See the garden of Alcinous, in Homer's Odyssey; those of Morgana, Alcina, and Armida, in the Italian poets: the gardens fair

> "Of Hesperus and his daughters three
> Who sing about the golden tree:"

and Proserpina's garden, and the Bower of Bliss in Spenser's Fairie Queene. The very mention of their names seems to embower one in leaves and blossoms.

It is a matter of some taste to arrange a bouquet of flowers judiciously; even in language, we have a finer idea of colours, when such are placed together as look well together in substance. Do we read of white, purple, red, and yellow flowers, they do not present to us so exquisite a picture, as if we read of yellow and purple, white and red. Their arrangement has been happily touched upon by some of our poets:

> ——————————" th' Azores send
> Their jessamine; her jessamine, remote
> Caffraia: foreigners from many lands,
> They form one social shade, as if convened
> By magic summons of th' Orphean lyre.
> Yet just arrangement, rarely brought to pass
> But by a master's hand disposing well
> The gay diversities of leaf and flower,
> Must lend its aid t' illustrate all their charms,
> And dress the regular, yet various scene.
> Plant behind plant aspiring, in the van
> The dwarfish; in the rear retired, but still
> Sublime above the rest, the statelier stand."
>
> Cowper.

> ——————————" tibi lilia plenis
> Ecce ferunt nymphæ calathis: tibi candida Nais,

Pallentes violas et summa papavera carpens,
Narcissum et florem jungit benè olentis anethi.
Tum casiâ, atque aliis intexens suavibus herbis,
Mollia luteolâ pingit vaccinia calthâ."

VIRGIL, Eclogue 2.

"Behold the nymphs bring thee lilies in full baskets: for thee fair Nais, cropping the pale violets and heads of poppies, joins the narcissus, and flower of sweet-smelling anise: then, interweaving them with cassia and other fragrant herbs, sets off the soft hyacinth with the saffron marygold."

DAVIDSON'S TRANSLATION.

Drayton runs riot on the subject: a nymph in his Muse's Elysium says,

"Here damask-roses, white and red,
Out of my lap first take I,
Which still shall run along the thread;
My chiefest flower this make I.
Amongst these roses in a row,
Next place I pinks in plenty,
These double-daisies then for show,
And will not this be dainty?
The pretty pansy then I'll tye
Like stones some chain inchasing;
And next to them, their near ally,
The purple violet placing.
The curious choice clove July-flower,
Whose kinds, hight the carnation,
For sweetness of most sovereign power
Shall help my wreath to fashion;
Whose sundry colours of one kind,
First from one root derived,
Them in their several suits I'll bind,
My garland so contrived:
A course of cowslips then I'll stick,
And here and there (though sparely)
The pleasant primrose down I'll prick,
Like pearls which will show rarely;
Then with these marygolds I'll make
My garland somewhat swelling,
These honeysuckles then I'll take,
Whose sweets shall help their smelling.

The lily and the fleur-de-lis,
For colour much contenting,
For that I them do only prize,
They are but poor in scenting;
The daffodil most dainty is
To match with these in meetness;
The columbine compared to this,
All much alike for sweetness:
These in their natures only are
Fit to emboss the border,
Therefore I'll take especial care
To place them in their order:
Sweet-williams, campions, sops-in-wine,
One by another neatly;
Thus have I made this wreath of mine,
And finished it featly."

DRAYTON.

" So did the maidens with their various flowers
Deck up their windows and make neat their bowers:
Using such cunning as they did dispose
The ruddy peony with the lighter rose,
The monkshood with the bugloss, and entwine
The white, the blue, the flesh-like columbine
With pinks, sweet-williams; that far off the eye
Could not the manner of their mixtures spy.'

W. BROWNE.

What is here said on the subject of arrangement is of course addressed to those who are unacquainted with botany; those who study that delightful science will, most probably, prefer a botanical arrangement, observing however to place the smaller plants of each division next the spectator, and thus proceeding gradually to the tallest and most distant; so that the several divisions will form stripes irregular in their width.

The exertions of Lamarcke and the Jussieus have now so improved the ancient and original method of arranging plants by their natural affinities to each other, that most of the young botanists have adopted it. The only work in which this truly scientific method is applied to all the plants growing wild in the British islands is Gray's Natural Arrangement; which also contains an Introduction to Botany in general, on a more extensive scale than Withering's, as it includes the explanation of all the new terms which have been lately introduced into botany by the cultivators of the natural system.

Although it is true that near London plants in general will not thrive so well as in a purer air, and that people in the country have usually some portion of ground to make a garden of, yet such persons as are condemned to a town life will do well to obtain whatever substitute for a garden may be in their power; for there is confessedly no greater folly than that of refusing all pleasure, because we cannot have all we desire. In Venice, where the nature of the place is such as to afford no garden ground, it is common to see the windows filled with pots, and they have a market for the sale of them. Those who can afford it, indeed, have gardens elsewhere; but by far the greater number are obliged to content themselves with a *portable garden.* A lover of flowers, who cannot have a garden or a green-house, will gladly cherish any thing that has the aspect of a green leaf;

————————" These serve him with a hint
That Nature lives: that sight-refreshing green
Is still the livery she delights to wear,
Though sickly samples of th' exuberant whole.
What are the casements lined with creeping herbs,
The prouder sashes fronted with a range
Of orange, myrtle, or the fragrant weed,
The Frenchman's darling*? Are they not all proofs,
That man immured in cities, still retains
His inborn, inextinguishable thirst

* Mignonette.

Of rural scenes, compensating his loss
By supplemental shifts, the best he may?"

COWPER.

With this passage, which brings us round to the direct object of this little work, it will be as well for me to conclude the preface. I am as fond of books as of flowers; but in all that regards authorship, I fear I am as little able to produce the one, as to create the others. I therefore hasten to the more mechanical part of my work, and to the kind aid of my quotations. I shall only add, if any body would like to have additional authority for the cultivation of a few domestic flowers, that Gray, with all his love of the grander features of nature, and all his nice sense of his own dignity, did not think it beneath him to supply the want of a larger garden with flower-pots in his windows, to look to them entirely himself, and to take them in, with all due tenderness, of an evening. See his delightful letters to his friends.

For a poetical translation of some quotations, of which there was before either no English version, or none that did justice to the original, as well as for some general corrections, &c. I am indebted to the assistance of a friend, whose kindness I most gratefully and somewhat proudly acknowledge, in sparing a few hours from his own important studies, to give this little volume some pretension to public notice.

Although no other flowers are considered in this work, but those usually grown in pots; yet this comprises a larger collection than most persons are likely to cultivate. They indeed who are much attached to the beauties of the vegetable tribes may add others not here mentioned, go very deep into the science of botany, and yet keep within the limits of a garden of pots. Some even of the most scientific botanists prefer a domestic garden of this kind.

For example, Richard Anthony Salisbury, Esq. the universally acknowledged head of our English botanists, no longer cultivates his former gardens at Chapel Allerton, Yorkshire, or at Mill Hill, Middlesex, but confines his attention to a choice collection of the most curious plants in pots, arranged in the yard of his house in Queen Street, Edgeware Road. In like manner, Messrs. Loddiges, nurserymen at Hackney, have a very large collection of hardy herbaceous plants, in small pots, set on beds of scoria, to keep the soil contained in them moist.

FLORA DOMESTICA.

FLORA DOMESTICA,

&c.

ADONIS.

RANUNCULACEÆ. POLYANDRIA POLYGYNIA.

Italian, adonio.—*French,* adonide; rose rubi; gouttes de sang [drops of blood]; aile de faisan [pheasant's-wing]; œil de perdrix [partridge's eye].—*Greek,* eranthemon [spring-flower].—*English,* adonis-flower; bird's eye; pheasant's eye; flos-adonis. The autumnal adonis is also called red maythes, red morocco; to which Gerarde adds may-weed, and red camomile. " Our London women," says he, " do call it rose-a-rubie."

This flower owes its classical name to Adonis, the favourite of Venus: some say its existence also; maintaining that it sprung from his blood, when dying. It is likely that the name arose from confounding it with the anemone, which it resembles. There are, however, other flowers which lay claim to this illustrious origin; the larkspur is one, but the claim is too weak to be generally allowed. Moschus has conferred this distinction on the rose. Others, again, trace its pedigree to the tears which Venus shed upon her lover's body; and Gerarde would persuade us that these tears gave birth to the Venice-mallow: but the anemone has pretty generally established her descent from both parents.—See *Anemone.*

The name of the beautiful huntsman, in his living capacity, however, applies well enough; for the Adonis is handsome and ruddy, and an enemy to the corn; but the flower is not so hardy as its godfather, and must be sheltered from the frosts of winter.

The Autumnal, or Common Adonis, has usually a red flower; but there is a variety of this species, of which the flowers are lemon-coloured. It is a native of most parts of the south of Europe; in Germany it grows wild among the corn; as it does, according to Gerarde, in the west of England. It is very common in some parts of Kent, particularly on the banks of the Medway,—a water-nymph, according to Spenser, famous for her flowers.

"Then came the bride, the lovely Medway came,
Clad in a vesture of unknowen geare,
And uncouth fashion, yet her well became,
That seemed like silver sprinkled here and there
With glittering spangs that did like stars appear,
And waved upon, like water chamelot,
To hide the metal, which yet every where
Bewrayed itself, to let men plainly wot
It was no mortal work, that seemed, and yet was not.

Her goodly locks adown her back did flow,
Unto her waist, with flowers bescattered,
The which ambrosial odours forth did throw
To all about, and all her shoulders spread
As a new spring: and likewise on her head
A chapelet of sundry flowers she wore,
From under which the dewy humour shed,
Did trickle down her hair, like to the hore
Congealed little drops which do the morn adore."

The Vernal Adonis [Fr. *hellebore d'Hippocrate*] is a perennial; and, as it does not flower the first year, it might be more convenient to purchase it at a nursery when in a state to flower, than to raise it at home. It may, however, be treated in the same manner as the Autumnal Adonis. It is a native of Switzerland, Germany, &c. It bears a large yellow flower, which blows about the end of March, or the beginning cf April.

The Apennine Adonis is very similar to the vernal, of which it is termed the sister; but it continues longer in flower than that species, which, true to the name it bears, comes and goes with the spring. The reader of poetry is aware that Adonis, after death, was supposed to spend his time alternately with Proserpine in the lower regions, and with Venus on earth.

> "Go, beloved Adonis, go,
> Year by year thus to and fro,
> Only privileged demigod!
> There was no such open road
> For Atrides; nor the great
> Ajax, chief infuriate;
> Not for Hector, noblest once
> Of his mother's twenty sons;
> Nor Patroclus; nor the boy
> That return'd from taken Troy *."

There is also a shrub Adonis, a native of the Cape of Good Hope.

The Autumnal Adonis is an annual, and the seeds sown in spring will flower in October. If some of the seeds are sown in September they will blow early in June. As the flowers open sooner or later in proportion to their exposure to the sun, a little attention to their arrangement will insure a longer succession of them. The seeds should be sown two or three in a pot, half an inch deep. During the severity of the winter, the pots should be housed; but in mild weather they should stand in the open air. In dry weather they should be occasionally, but sparingly, watered, just enough to preserve them from drought.

* See the Translations from Theocritus, in Hunt's Foliage.

AFRICAN LILY.

AGAPANTHUS.

HEMEROCALLIDEÆ. HEXANDRIA MONOGYNIA.

The botanical name of this flower is from the Greek, and signifies a delightful flower.

This Lily is a native of the Cape of Good Hope: it is of a bright blue colour; very showy and elegant. The flowers blow about the end of August, and will frequently preserve their beauty till the spring.

It is increased by offsets, which come out from the sides of the old plants, and may be taken off at the latter end of June; at which time the plant is in its most dormant state. It should be turned out of the pot, and the earth carefully cleared away, that the fibres of the offsets may be the better distinguished: and these must be carefully separated from those of the old root. Where they adhere so closely as not to be otherwise parted, they must be cut off with a knife; great care being taken not to wound or break the bulb, either of the offset or of the parent plant.

When these are parted, they should be planted, each in a separate pot filled with light kitchen garden earth, and placed in a shady situation, where they may enjoy the morning sun; a little water should be given to them twice a week, if the weather be dry; but they must not have much, especially at this season, when they are almost inactive; for as the roots are fleshy and succulent, they are apt to rot with too much moisture. In about five weeks the offsets will have put out new roots: they may then be removed to a more sunny situation, and may have a little more water; but still not plentifully. In September they will put out their flower-stalks, and towards the end of the month, the flowers will begin to open; when, unless the weather be very fine, they should be housed, that they may not be injured by too much wet, or by frost; but they must be allowed as much fresh air as possible. During the winter they may have a little water once a week in mild weather, but none in frost. This flower must be watered only at the roots.

ALMOND TREE.

AMYGDALA.

ROSACEÆ. ICOSANDRIA MONOGYNIA.

THE Almond-tree! the lofty Almond-tree a potted plant! the Almond tree, to which Spenser, in an exquisite passage, likens the plume of Prince Arthur:

> " Upon the top of all, his lofty crest,
> A bunch of hairs discolour'd diversly,
> With sprinkled pearl and gold full richly drest,
> Did shake, and seem'd to dance for jollity.
> Like to an almond-tree ymounted high
> On top of green Selinis all alone,
> With blossoms brave bedecked daintily,
> Whose tender locks do tremble every one
> At every little breath that under heaven is blown."

No, it is not this immortal Almond-tree that is to be moved at pleasure from the garden to a room or balcony; but a Russian cousin, the Bobownik, Dikii Persik, or Calmyzkii Orech [Calmuck almond]; but called by the Calmucks themselves, Charun Orak, a young Tartar of humble growth, though emulating his great relation in the elegance of his apparel. He is called the Dwarf Almond tree; and is worthy to have derived his name from the transformation of some dwarf in a fairy tale into a tree. In April the young shoots of this tree are covered with blossoms of a beautiful blush-colour; and the leaves are sometimes five inches long. It will bear the open air, and, when the weather is dry, should be watered every evening. The young suckers from the roots must be taken off every year, or they will starve the parent plant: they may be planted in February or October, and should be placed in the shade till they have taken root. The fruit of this shrub is about the size of a hazel-nut, and has the taste of the peach-kernel.

Plutarch mentions a great drinker of wine, who, by the use of bitter almonds, used to escape being intoxicated. The Italians, upon their favourite modern principle of contra stimulants, suppose this very likely; and so it may be; but it need not be added, that to tamper in this manner with diseases seems very dangerous.

ALOE.

ASPHODELEÆ. HEXANDRIA MONOGYNIA.

The derivation of this name is uncertain. Beginning with the syllable Al, it is, perhaps, of Arabian origin; especially as the plant is much venerated in the East. In the Hebrew, a cognate language, it is called ahalah: some derive Aloes from the Greek *als* [the sea]; others from the Latin, *adolendo;* but this can only refer to the Aloe-wood, which is used in sacrifices for its fragrance. On the whole, it is probable the name was first applied to the Aloe-wood, and hence transferred to the common Aloes, on account of their bitterness. Its medicinal virtues were made known to us by Dioscorides, the physician of Cleopatra; and it is also mentioned by Plutarch. The name Aloe is retained by all the European nations.

From the specimens we are in the habit of seeing in this country, we should be inclined to think that the utility of the Aloe far surpassed its beauty, and to rank it, as a vegetable, with the camel and the elephant in animal life. Like the larger animals, it is confined to hot, or comparatively uncivilised countries. Its appearance, which resembles a collection of huge leathern claws, armed with prickles, is very formidable; and even the smaller species have a sort of monstrosity of size in their parts, though small as a whole. But notwithstanding the extraordinary utility of the Aloe, those who have seen it in its native country, and in full flower, describe it as scarcely less remarkable for elegance and beauty. The larger and more useful kinds appear to be also the most beautiful. Rousseau uses the epithet beautiful, in speaking of the great American Aloe, or Agave.

"Nature seems to have treated the Africans and Asiatics as barbarians," says St. Pierre, in speaking of the Aloe, "in having given them these at once magnificent, yet monstrous vegetables; and to have dealt with us as beings capable of sensibility and society. Oh, when shall I breathe the perfume of the honeysuckle?—again repose myself upon a carpet of milk-weed, saffron, and blue-bells, the food of our lowing herds? and once more hear Aurora welcomed with the songs of the labourer, blessed with freedom and content*?"

The kind chiefly used in medicine is the Barbadoes Aloe, the preparations from which are eminent for the nauseousness of their bitter. "As bitter as aloes," is a proverbial phrase. It is a common practice with our fair countrywomen to avail themselves of this bitterness in the Aloe, when weaning their children; applying it to the bosom to induce them to refuse it; but this is surely a more objectionable deceit than that by which they are allured to swallow nauseous drugs.

* St. Pierre's Voyage to the Isle of France.

"Cosi all' egro fanciul porgiamo aspersi
Di soavi licor gli orli del vaso:
Succhi amari, ingannato, intanto ei beve;
E dall' inganno suo vita riceve."

TASSO.

So we (if children young diseased we find)
Anoint with sweets the vessel's foremost parts,
To make them taste the potions sharp we give:
They drink, deceived; and so deceived, they live.

FAIRFAX'S TRANSLATION.

It seems strange that any thing but the most imperative necessity should induce a mother to use any means which can render her an object of disgust to her child.

The most remarkable of the Aloe tribe is the great American Aloe, named by botanists Agave, which name is derived from the Greek, and signifies admirable, or glorious: called by the French *aloe en arbre* [tree aloe], and also *pitte.* The natural order in which it should be arranged is uncertain. Bernard Jussieu placed it with the Narcissi, and Anthony Jussieu with the Bromeliaceæ. It is a native of all the southern parts of America. "The stem generally rises upwards of twenty feet high, and branches out on every side towards the top, so as to form a kind of pyramid. The slender shoots are garnished with greenish yellow flowers, which come out in thick clusters at every joint, and continue long in beauty; a succession of new flowers being produced for near three months in favourable seasons, if the plant is protected from the autumnal cold. The elegance of the flower, and the rarity of its appearance in our cold climate, render it an object of such general curiosity, that the gardener who possesses the plant announces it in the public papers, and builds a platform round it for the accommodation of the spectators. The popular opinions, that the aloe flowers but once in a century, and that its blooming is attended with a noise like the report of a cannon, are equally without foundation. The fact is, that the time which this plant takes to come to perfection varies with the climate. In hot countries, where they grow fast, and expand many leaves every season, they will flower in a few years; but in colder climates, where their growth is slow, they will be much longer in arriving at perfection. The leaves of the American Aloe are five or six feet long, from six to nine inches broad, and three or four thick *."

Millar mentions one of these plants in the garden of the King of Prussia, that was forty feet high; another in the royal garden at Friedricksberg in Denmark, two-and-twenty feet high, which had nineteen branches, bearing four thousand flowers; and a third in the botanic garden at Cambridge, which, at sixty years of age, had never borne flowers. He specifies some others, remarkable for the number of their flowers, but does not mention the age of any one at the time of flowering.

"With us," says Rousseau, "the term of its life is uncertain; and after having flowered, it produces a number of offsets, and dies."

A kind of soap is prepared from the leaves of this Aloe, and the leaves themselves are used for scouring floors, pewter, &c.; and their epidermis is serviceable to literature as a material for writing upon. The following extract from Wood's Zoography will give some idea of the general utility of this extraordinary plant:—

"The Mahometans respect the Aloe as a plant of a superior nature. In Egypt it may be said to bear some share in their religious ceremonies; since whoever returns from a pilgrimage to Mecca hangs it over his street-door as a proof of his having performed that holy journey. The

* Wood's Zoography, vol. iii.

superstitious Egyptians believe that this plant hinders evil spirits and apparitions from entering the house; and on this account, whoever walks the streets in Cairo, will find it over the doors of both Christians and Jews.

" The leaves of the different specimens of Aloe, as well as the Agave, are highly serviceable to the natives of the countries where they grow. The negroes in Senegal make excellent ropes of them, which are not liable to rot in water; and of two kinds mentioned by Sir Hans Sloane, one is manufactured into fishing-lines, bow-strings, stockings, and hammocks; while the other has leaves, which, like those of the wild pine and the banana, hold rain-water, and thus afford a valuable refreshment to travellers in hot climates. The poor in Mexico derive almost every necessary of life from a species of Aloe. Besides making excellent hedges for their fields, its trunk serves instead of beams for the roofs of their houses, and its leaves supply the place of tiles. From these they obtain paper, thread, needles, clothing, shoes, stockings, and cordage; from the juice they make wine, honey, sugar, and vinegar."

Such of the Aloes as do not require a stove will bear the open air, in our climate, from the end of March to the end of September. During the winter they should be watered about once in a month; in the summer, when the weather is dry, once in a week or ten days; but when there is much rain, they should be sheltered from it, or they will be apt to rot. If the weather be mild, they may be placed where they may receive the fresh air in the day-time for a month after they are housed; after that the windows should be closed. They should not be put into large pots, but should be removed into fresh earth every year, which should be done in July. As much of the earth should be shaken away as possible, the roots opened with the fingers, and such as are decayed taken off; but great care must be taken not to break or wound those which are young and fresh. Water them gently when newly planted, place them in the shade for three weeks, and if the weather is hot and dry, water them in a similar manner once or twice a week. Most of the species may at this time be increased by offsets, which should be planted in very small pots; and if, in taking off the suckers, you find them very moist where they are broken from the mother-root, they should lie in a dry shady place for a week before they are planted. When planted, treat them like the old plants. Such kinds as do not afford plenty of offsets may generally be propagated by taking off some of the under leaves, laying them to dry for ten days or a fortnight, and planting them, putting that part of the leaf which adhered to the old plant about an inch or an inch and a half into the earth. This should be done in June.

There are few things, I believe, more venerable, more eloquently impressive in their antiquity, than an old tree. The ruins of an old and noble edifice, of which every shattered fragment, every gaping cranny, complains of the destructive hand of time, is young and modern in our eyes, compared with that which still survives its touch,—the old ivy, that still, with every succeeding year, moves slowly on, knitting its creeping stalks into every crevice, and carrying its broad leaves up to the very summit. What can be more venerable than the far-spreading roots of an old elm or oak tree, veining the earth with wood! Cross but that little piece of wood, called the wilderness, leading from Hampstead towards North End, where the intermingled roots are visible at every step, casing the earth in impenetrable armour, and forming a natural pavement, apparently as old as time itself—can all the antiquities of Egypt command a greater reverence?

The larger species of Aloe, from the immensity of its

size, and the known slowness of its growth, must speak the same impressive language. Mr. Campbell has put it in a noble attitude for the occasion:

> "Rocks sublime
> To human art a sportive semblance bore,
> And yellow lichens colour'd all the clime
> Like moonlight battlements, and towers decay'd by time.
> But high in amphitheatre above,
> His arms the everlasting aloes threw."
>
> GERTRUDE OF WYOMING.

The Abbé la Pluche gives an interesting account of the uses of the Chinese Aloe, commonly called Wood-aloes, or Aloes-wood: from whence, as has been supposed, the name of aloe has been transferred to the common species.

"This Aloe," says he, "is as tall as the olive-tree, and of much the same shape: there are three sorts of wood contained under its bark; the first is black, compact, and heavy; the second swarthy, and as light as touchwood; the third, which lies near the heart, diffuses a powerful fragrance. The first is known by the name of eagle-wood, and is a scarce commodity; the second, calembouc-wood, which is transported into Europe, where it is highly esteemed as an excellent drug; it burns like wax, and, when thrown into the fire, has an aromatic odour. The third, which is the heart, and called calambac, or tambac wood, is a more valuable commodity in the Indies than gold itself. It is used for perfuming the clothes and the apartments of persons of distinction; and is a specific medicine for persons affected with fainting-fits, or with the palsy. The Indians, likewise, set their most costly jewels in this wood. The leaves of this tree are sometimes used instead of slates for roofing houses; are manufactured into dishes and plates, and, when well dried, are fit to be brought to table. If stripped betimes of their nerves and fibres, they are used as hemp, and manufactured into a thread. Of the points, with which the branches abound, are made nails, darts, and awls. The Indians pierce holes in their ears with the last, when they propose to honour the devil with some peculiar testimonies of their devotion. If any orifice or aperture be made in this tree by cutting off any of its buds, a sweet vinous liquor effuses in abundance from the wound, which proves an agreeable liquor to drink when fresh, and in process of time becomes an excellent vinegar. The wood of the branches is very agreeable to the taste, and has something of the flavour of a candied citron. The roots themselves are of service, and are frequently converted into ropes. To conclude, a whole family may subsist on, reside in, and be decently clothed by, one of these Aloes."

The common writing-paper in Cochin-China is made from the bark of this tree; of which the botanical name is aquilaria, from *aquila*, an eagle, so named because it grows in lofty places; and from its bitter taste, also termed Wood-aloes.

Chaucer notices both the fragrance and the bitterness of the Aloe-wood:

"The woful teris that thei letin fal
As bittir werin, out of teris kinde,
For paine, as is ligne aloes, or gal."
TROILUS AND CRESEIDE, book iv.

"My chambir is strowed with mirre and insence,
With sote savoring aloes and sinnamone,
Brething an aromatike redolence."
REMEDIE OF LOVE.

The great antiquity of the use of Wood-aloe as a perfume is shown by the Bible: "All thy garments," says a passage in the Psalms, "smell of myrrh, and aloes, and cassia:"

and Solomon, addressing the object of his love, says, "thy plants are an orchard of pomegranates, with pleasant fruits; camphire, with spikenard; spikenard and saffron; calamus and cinnamon, with all trees of frankincense; myrrh and aloes, with all the chief spices: a fountain of gardens, a well of living waters, and streams from Lebanon:" upon which, the object of his love, as if in an enthusiasm of delight at his speaking so of the place she lives in, beautifully exclaims, "Awake, O north wind; and come, thou south; blow upon my garden, that the spices thereof may come out. Let my beloved come into his garden, and eat his pleasant fruits."

AMARANTH.

AMARANTHUS.

AMARANTHACEÆ. MONŒCIA PENTANDRIA.

Italian, amaranto, fior veluto [velvet-flower]; maraviglie di Spagna [the Spanish wonder].—*French*, amaranthe; passe-velours [pass-velvet]; fleur d'amour [love-flower] —*English*, amaranth; flower-gentle; velvet-flower. The botanical name is derived from the Greek, and signifies unfading.

The species of Amaranth most cultivated in English gardens are the Two-coloured Amaranth, which flowers late in the autumn, with purple and crimson flowers;—the Three-coloured Amaranth, with variegated flowers, which continue to blow from June to September (Fr. *fleur de jalousie*, jealous-flower; in Spanish and Portuguese called *papagayo*, the parrot); "there is not," says Millar, "a handsomer plant than this in its full lustre;"—the Prince's-feather Amaranth (amar. hypochondriacus), which also varies in colour, and which flowers at the same time;—the Spreading or Bloody Amaranth, with flowers of a red purple, blowing from June to September;—the Pendulous Amaranth, or Love-lies-bleeding, (Fr. *discipline des religieuses,* the nuns' whipping rope), with flowers of a red purple, blowing in August and September;—the Cock's comb, or Crested Amaranth [Celosia in pentandria monogynia], of which the flowers are red, purple, white, yellow, or variegated, flowering in July and August;—and the Globe Amaranth [Gomphrena in pentandria digynia; but, like Celosia, still belonging to the same natural family of Amaranthaceæ], of which there are several varieties, white, purple, striped, &c. The purple resembles clover raised to an intense pitch of colour, and sprinkled with grains of gold. The flowers, gathered when full grown, and dried in the shade, will preserve their beauty for years, particularly if they are not exposed to the sun. A friend of the writer's possesses some Amaranths, both purple and yellow, which he has had by him for several years, enclosed with some locks of hair in a little marble urn. They look as vivid as if they were put in yesterday; and it may be added, that they are particularly suited to their situation. They remind us of Milton's use of the Amaranth, when speaking of the multitude of angels assembled before the Deity:

——————— "to the ground
With solemn adoration down they cast
Their crowns inwove with amaranth and gold;
Immortal amaranth, a flower which once
In Paradise, fast by the tree of life,
Began to bloom, but soon for man's offence
To heaven removed, where first it grew, there grows
And flowers aloft, shading the fount of life,
And where the river of bliss through midst of heaven
Rolls o'er Elysian flowers her amber stream;
With these that never fade, the spirits elect

Bind their resplendent locks enwreathed with beams;
Now in loose garlands thick thrown off, the bright
Pavement, that like a sea of jasper shone,
Impurpled with celestial roses smiled."

The following occurs in Shelley's Rosalind and Helen:

"Whose sad inhabitants each year would come,
With willing steps climbing that rugged height,
And hang long locks of hair, and garlands bound
With amaranth flowers, which, in the clime's despite,
Filled the frore air with unaccustomed light.
Such flowers as in the wintery memory bloom
Of one friend left, adorned that frozen tomb."

In Portugal, and other warm countries, the churches are, in winter, adorned with the Globe Amaranth. Cowley and Rapin, in their Latin poems on plants and gardens, make honourable mention of the Amaranth; but the translations of those poems are too unworthy of their originals to admit of quotation, and a friend who would have supplied me with better is on a distant journey.

The Cock's comb Amaranth is a very showy and remarkable plant. The appellation was given it from the form of its crested head of flowers resembling the comb of a cock. Sometimes the heads are divided like a plume of feathers. It is said that in Japan these crests or heads of flowers are often a foot in length and in breadth, and extremely beautiful. The colour of the scarlet varieties is highly brilliant.

The Amaranths are all annual, must be raised in a hot-bed, and may be had from a nursery when strong enough to bear removal, which, for the last three kinds, will not be earlier than the middle of June: the others may be placed abroad earlier. In dry weather they should be watered every evening. Such flowers as are intended to be preserved should be cut before they run to seed; and should be observed daily after they are blown, that they may be taken in full beauty.

The Amaranth is recommended, among other flowers, as a food for bees:

" Il timo e l' amaranto
Dei trapiantare ancora, e quell' altr' erbe
Che danno a questa greggia amabil cibo."

LE API DEL RUCELLAI.

Thyme and the amaranth
Also transplant, and all such other herbs
As yield the winged flock a food they love.

Moore speaks of them as being used for the hair, a purpose for which they are peculiarly well adapted:

" Amaranths such as crown the maids
That wander through Zamara's shades *."

From a passage in Don Quixote one may suppose that Amaranths were sometimes worn by the Spanish ladies in the time of Cervantes; but the chief value of such passages consists in showing us the probable taste of the author. It is where he speaks of a set of ladies and gentlemen who were amusing themselves by playing shepherds and shepherdesses in the woods, and who had hung some green nets across the trees. And as he (Don Quixote) was going to pass forward and break through all (he took it for the work of enchanters) "unexpectedly from among some trees two most beautiful shepherdesses presented themselves before him: at least they were clad like shepherdesses, except that their waistcoats and petticoats were of fine brocade, their habits were of rich gold tabby, their hair, which for brightness might come in competition with the rays of the sun, hanging loose about their shoulders, and their heads crowned with garlands of green laurel and red flower-gentles interwoven." The delicate and sunny-coloured bay leaves of the south, and the red or

* " The people of the Batta country, in Sumatra, or Zamara, when not engaged in war, lead an idle inactive life, passing the day in playing on a kind of flute, crowned with garlands of flowers, among which the Globe Amaranth, a native of the country, mostly prevails."

purple Amaranth, interwoven, would make a beautiful mixture, especially as the Amaranth is deficient in leaves.

One of the most popular species of the Amaranth is the Love-lies-bleeding. The origin of this name is not generally known; unless we are to suppose it christened by the daughter of O'Connor, in her tender lamentations over the tomb of Connocht Moran:

"A hero's bride! this desert bower,
It ill befits thy gentle breeding:
And wherefore dost thou love this flower
To call—my love-lies bleeding?

This purple flower my tears have nursed;
A hero's blood supplied its bloom:
I love it, for it was the first
That grew on Connocht-Moran's tomb."

The Amaranths are chiefly natives of America, and very few are supposed to grow naturally in Europe; yet Sir W. Jones speaks of them as if growing wild in Wales:

"Fair Tivy, how sweet are thy waves gently flowing,
Thy wild oaken woods, and green eglantine bowers,
Thy banks with the blush-rose and amaranth glowing
While friendship and mirth claim their labourless hours!"

ANDROMEDA.

ERICINEÆ. DECANDRIA MONOGYNIA.

Marsh cistus; wild rosemary; poley-mountain; moon-wort; marsh holy-rose.

THIS plant was named by Linnæus, from the daughter of Cepheus and Cassiope, who was exposed at the waterside, and rescued from the sea-monster by Perseus. Thus a name in botany, especially in the works of this great and illustrious naturalist, is often made to tell two stories—that of its classical prototype and of its own nature.

The Marsh Andromeda, which is a native of America and many parts of Europe, is also a plant of our own;

growing wild in most of our northern counties, as well as in the Lowlands of Scotland. It is an elegant little shrub, with pink flowers, which begin to open toward the end of May.

This is the species of Andromeda the most desirable for home-cultivation; but there are many others, of which two or three are evergreens; as the willow-leaved and the box-leaved Andromedas. They will all bear the open air. In dry summer weather they will require water every evening; if the weather be very hot, they may be watered in the morning also.

ANEMONE.

RANUNCULACEÆ. POLYANDRIA POLYGYNIA.

Anemone, from the Greek, anemos, wind: some say because the flower opens only when the wind blows; others, because it grows in situations much exposed to the wind.—*French,* Anemone, l'herbe au vent [wind herb].

To do justice to every species of the Anemone, it would be necessary to write a volume upon that subject alone; but it will suffice for the present purpose to speak of the kinds most desirable.

The Anemones are natives of the East, from whence their roots were originally brought; but they have been so much improved by culture, as to take a high rank among the ornaments of our gardens in the spring. As they do not blow the first year, it will be more convenient to purchase the plants from a nursery than to rear them at home: on another account also, it will be better; for they vary so much, that it is impossible to secure the handsomest kinds by the seed; and, when in flower, they may be selected according to the taste of the purchaser. They should be sheltered from frost and heavy rains: light showers will refresh them, and in dry weather they should be watered every evening, but very gently. When the roots are once obtained, they may be increased by parting.

The Narrow-leaved Garden Anemone grows wild in the Levant. In the islands of the Archipelago the borders of the fields are covered with it in almost every variety of colour; but these are single; culture has made them double.

Of the double varieties of this species there are nearly two hundred. To be a fine one, a double Anemone should have a strong upright stem, about nine inches high; the flower should be from two to three inches in diameter: the outer petals should be firm, horizontal, unless they turn up a little at the end, and the smaller petals within these should lie gracefully one over the other. The plain colours should be brilliant, the variegated clear and distinct.

The Broad-leaved Garden Anemone is found wild with single flowers, in Germany, Italy, and Provence; the single varieties are sometimes called Star-Anemones: they are very numerous, as are also the double varieties, of which the most remarkable are the great double Anemone of Constantinople, or Spanish marygold, the great double Orange-tawney, the double Anemone of Cyprus, and the double Persian Anemone.

There is a species called the Wood-Anemone, which grows in the woods and hedges in most parts of Europe. In March, April, and May, many of our woods are almost covered with these flowers, which expand in clear weather, and look towards the sun; but in the evening, and in wet weather, close and droop their heads. When the Wood-Anemone becomes double, it is cultivated by the gardeners; and were the same pains taken with this as with the foreign Anemones, it would probably become valuable.

Anemone roots may be planted towards the end of September, and again a month later; some plant a third set about Christmas. The first planted will begin to flower early in April, and continue for three or four weeks; the others will follow in succession. As soon as the leaves decay, which of those first planted, will be in June, the roots should be taken up, the decayed parts and the earth cleared away; and, having been dried in the shade, they should be put in some secure place, where they may be perfectly dry, and particularly where mice, &c. cannot find access to them. This opportunity may be taken to part the roots for increase; and provided each part has a good eye or bud, it will grow and flower; but they will not flower so strong if parted small. The roots will be weakened, if suffered to remain long in the earth after the leaves decay. They will keep out of the earth for two, or even three years, and grow when planted. The single, or Poppy Anemone, will, in mild seasons, blow throughout the winter.

Earth proper for the Anemone may be procured from a nursery; the roots may be planted in pots five inches wide; the earth an inch and a half deep over the top of the roots, and the eye of the root upwards. They must be kept moderately moist, shaded from the noon-day sun, and exposed to that of the morning. In the winter they should be placed under shelter, but should have plenty of fresh air, when not frosty.

The Abbé la Pluche relates a curious anecdote of M. Bachelier, a Parisian florist, who, having imported some very beautiful species of the Anemone from the East Indies to Paris, kept them to himself in so miserly a manner, that for ten successive years he never would give to any friend or relation whomsoever the least fibre of a double Anemone, or the root of one single one. A counsellor of the parliament, vexed to see one man hoard up for himself a benefit which nature intended to be common to all, paid him a visit at his country-house, and, in walking round the garden, when he came to a bed of his Anemones, which were at that time in seed, artfully let his robe fall upon them; by which device, he swept off a considerable number of the little grains, which stuck fast to it. His servant, whom he had purposely instructed, dexterously wrapped them up in a moment, without exciting any attention. The counsellor a short time after communicated to his friends the success of his project; and by their participation of his innocent theft, the flower became generally known.

Rapin, in his poem on gardens, ascribes the birth of the Anemone to the jealousy of Flora; who fearing that the incomparable beauty of a Grecian nymph would win from her the love of her husband Zephyr, transformed her into this flower. But to this tale he adds an account better authorised, of the Anemone having sprung from the blood of Adonis and the tears of Venus shed over his body; and it is but common justice to Flora to observe that this is the generally received opinion of the origin of the Anemone. Cowley gives it this parentage, in his poem on plants. Ovid describes Venus lamenting over the bleeding body of her lover, whose memory and her own grief she resolves to perpetuate by changing his blood to a flower; but less poetically than some others: he substitutes nectar for the tears of Venus; not even hinting that the said nectar was the tears of the goddess.

" But be thy blood a flower. Had Proserpine
The power to change a nymph to mint?—Is mine
Inferior? or will any envy me
For such a change? Thus having utter'd, she
Pour'd nectar on it, of a fragrant smell;
Sprinkled therewith, the blood began to swell,
Like shining bubbles that from drops ascend;
And ere an hour was fully at an end,
From thence a flower, alike in colour, rose,
Such as those trees produce, whose fruits enclose

> Within the limber rind their purple grains;
> And yet the beauty but awhile remains;
> For those light-hanging leaves, infirmly placed,
> The winds, that blow on all things, quickly blast."
>
> SANDYS'S OVID, book X.

The Greek poet, Bion, in his epitaph on Adonis, makes the Anemone the offspring of the goddess's tears.

Mr. Hor. Smith, in his poem of Amarynthus, supports the first reason for naming this flower the wind-flower—that it never opens but when the wind blows:

> "And then I gather'd rushes, and began
> To weave a garland for you, intertwined
> With violets, hepaticas, primroses,
> And coy Anemone, that ne'er uncloses
> Her lips until they 're blown on by the wind."
>
> AMARYNTHUS, p. 46.

It seems more usual, as well as in character, for the presence of the sun to unclose the lips of the Anemone, which commonly close when he withdraws; but when he shines clear,

> "Then thickly strewn in woodland bowers,
> Anemonies their stars unfold."

Sir W. Jones has translated an ode from the Turkish of Mesihi, in which the author celebrates several of the more sweet or splendid flowers:

> "See! yon anemones their leaves unfold,
> With rubies flaming, and with living gold."

"The sweetness of the bower has made the air so fragrant, that the dew, before it falls, is changed into rose water."

> "The dew-drops, sweeten'd by the musky gale,
> Are changed to essence ere they reach the dale."

The only poetical allusion, which I have met with, to

the fragility of the Anemone, is in the poems of Sir W. Jones:

> "Youth, like a thin anemone, displays
> His silken leaf, and in a morn decays."

ANTHOLYZA.

IRIDEÆ. TRIANDRIA MONOGYNIA.

The name of this flower is from two Greek words, signifying a flower and madness. Why they are so applied I do not know, unless it has been used in hydrophobia.

The Antholyzas being chiefly from warmer countries, will not bear the open air in this: they are usually kept within doors from October, until they have ceased flowering; when, if it is intended to save the seeds, they are set abroad to perfect them; but the better mode of raising them in private gardens is to part the offsets from the bulbs, which furnish them in plenty. Those raised from seed do not flower till the third year. The best time to plant the roots is in August; they should be housed at the end of September, and will continue growing all the winter. In April, or early in May, the flowers appear: when these and the leaves have decayed, the bulbs should be taken up, dried in the shade, and cleaned, and preserved as directed for other bulbs. In August they may be replanted: the offsets may be planted three or four in a pot, the first year; the second, they should be separated to flower. In winter, they should be gently watered once or twice a week; in the spring, they will require it oftener, perhaps every evening, but sparingly.

The principal species are the Plaited-leaved Antholyza, with red flowers; the Scarlet-flowered, which is very beautiful; the Broad-leaved, which has also scarlet flowers;

and the Red-flowered [or Antholyza Meriana, *Fr.* la merianelle, so named by Dr. Trew, from Sybilla Merian, the celebrated female Dutch botanist; but placed by some in the genus Gladiolus; and by others in Watsonia], of which the flowers are of a copper-red colour outside, and of a deeper red within. They are all handsome plants; having, in addition to the beauty of their flowers, large dark green leaves, some of them a foot in length; they are natives of the Cape of Good Hope.

ANTHYLLIS.

LEGUMINOSÆ. DIADELPHIA DECANDRIA.

Kidney-vetch; ladies-finger; Jupiter's beard; silver bush. The name Anthyllis is derived from the Greek, and signifies a downy-flower; from the down on its leaves.—*French,* barbe de Jupiter [Jupiter's beard].—*Italian,* barba di Giove, signifying the same.

The Silvery Anthyllis, which is the only species necessary to mention here, is so called from the whiteness of its leaves: it is a handsome shrub, bearing yellow flowers which blow in June. This Anthyllis is a native of France, Spain, Italy, Portugal, and the East. It must be sheltered in winter; but the more air it enjoys in mild weather, the better it will thrive: in dry weather it should be gently watered every evening; in winter once a week will suffice.

Cuttings planted in any of the summer months in a pot of light earth, and placed in the shade, will take root, and may then be treated in every respect as the older plants.

Linnæus observes of the common Anthyllis, that the colour of the flowers varies with that of the soil: in Poland, where the soil is a red calcareous clay, the flowers are red: in Gothland, where the soil is white, the flowers are the same: here they are yellow.

ANTIRRHINUM.

PERSONEÆ. DIDYNAMIA ANGIOSPERMIA.

Toad-flax; snap-dragon; from the resemblance of its flowers to an open mouth.—*French,* mufle de veau.—*Italian,* antirrino.

THESE flowers are many of them large and handsome, but some persons consider them coarse; which, indeed, is the case with many of the most splendid flowers, as the hollyhock and the sun-flower. They are, however, very magnificent, particularly the great snap-dragon, or calve's snout; called by the French, *le muflier commun; mouron violet* [violet pimpernel]; *œil de chat* [cat's eye]; *gueule de lion;* &c. The flowers of this species are red, white, purple, yellow, or a combination of any two of these colours. They are single or double. It is a native of the south of Europe, and blows in June and July. The Russians express an oil from the seeds, little inferior to the oil of olives. This species is increased by cuttings planted in the summer in a dry soil: and this and the following are the kinds most commonly cultivated in gardens:

2. The three-leaved; Valentia and Sicily; purple or yellow; July and August.
3. The branching; Spain; yellow; May and June.
4. The violet-flowered; France and Italy.
5. The many-stalked; Sicily and the Levant; yellow; July.
6. The hairy; Spain; yellow; July.
7. The common yellow; Europe; June to August.
8. The brown-leaved; Siberia, Piedmont, &c.; yellow.
9. The purple, or Vesuvian; July to September.
10. The Montpelier; sweet-scented; blue; June to the end of autumn.
11. The dark-flowered; Gibraltar; flowers most of the summer.
12. The Alpine; very elegant; a fine violet-colour, with a rich gold-colour in the middle; many growing close together; all the summer.

2, 3, 4, 5, 6, are annual plants, and must be increased by seeds, which may be sown in the spring;—or in autumn, sheltering them in the winter; with the exception of the last, which should be sown in March, and will require no shelter. 3, 4, in five-inch pots: 6, three or four seeds in an eight-inch pot.

7, 8, 9, 10, are perennial plants; they may be sown as the last mentioned, in spring, or in autumn; they will require shelter from hard frost. The two last may also be increased by parting the root in autumn. The common-yellow is an indigenous plant, and if in a tolerably dry soil, will bear frost itself: a little straw over the roots will suffice for 8. In Worcestershire the common yellow toad-flax is called butter-and-eggs. It has leaves somewhat similar to flax, and on that account is named toad-flax, flax-weed, and wild flax. Its juice, mixed with milk, is used as a poison for flies; and water distilled from it is said to remove inflammation in the eyes.

11, 12, may be increased by cuttings, planted in the summer in a light unmanured soil. They must be removed into the house in October, and brought out again about the end of April, or early in May.

ARBOR-VITÆ.

THUJA.

CONIFERÆ. MONŒCIA ADELPHIA.

The origin of this name, which signifies the tree of life, does not appear, though it seems to have reference to the tree mentioned in the book of Genesis.—*French,* l'arbre de vie; cedre Americain [American cedar].—*Italian,* albero di vita.

The Arbor-vitæ is a native of Siberia and Canada, where it is very plentiful. Being the strongest wood in Canada, it is there used for enclosures and palisades, for boats, and the floors of rooms. It is reckoned one of the best woods for the use of the lime-kiln; and besoms made of its branches are carried over Canada by the Indians for sale. When fresh, they have a very agreeable scent, which is perceptible in houses swept with them. The leaves have medicinal properties. In England the wood is used for bowls, boxes, cups, &c.

This tree is sometimes called the white cedar. It begins to flower about May. A young plant may be procured from a nursery as soon as its education is so far advanced that it may be introduced to the world with propriety. It will thrive well in a pot for many years: but the best species for this purpose is the Chinese Arbor-vitæ, which does not grow too large for a pot. It will bear our climate in all its seasons, only requiring to be watered occasionally in dry weather.

ARBUTUS.

ERICINEÆ. DECANDRIA MONOGYNIA.

Strawberry-tree.—*French,* le fraisier en arbre, l'arbre à fraises, both similar to the common English name: the fruit is called arbouse, arboise, or arboust.—*Italian,* arbuto, albatro, albaro, corbezzolo, from the fruit called corbezzola. By Pliny the fruit is called unedo.

THIS is called the strawberry-tree, from the resemblance of its fruit to a strawberry. Although it attains a considerable size, it is frequently grown in pots, and will bear transplanting very well. For this operation, April is the most favourable time; the cultivator taking care to preserve the earth about the roots, and to shade them from the mid-day sun, when newly planted.

As the leaves of the Arbutus remain all the winter, and in spring are pushed off by the shooting of new ones, the tree is always clothed. In June the young leaves are extremely beautiful; in October and November it is one of the most ornamental trees we have; the blossoms of the present, and the ripe fruit of the former year, both adorning it at the same time. There is an Arbutus now in the garden (in October) before my window, more lovely than I can find language to express. When other trees are losing their beauty, this is in its fullest perfection; and realises the exuberant fiction of the poets,—bearing at once flowers and fruit:

"There is continual spring, and harvest there
Continual, both meeting at one time;
For both the boughs do laughing blossoms bear,
And with fresh colours deck the wanton prime,
And eke at once the heavy trees they climb,
Which seem to labour under their fruit's load:
The whiles the joyous birds make their pastime
Amongst the shady leaves, their sweet abode,
And their true loves without suspicion, tell abroad."

SPENSER'S FAERIE QUEENE.

——— "Great Spring, before,
Greened all the year: and fruits and blossoms blushed
In social sweetness on the self-same bough."

THOMSON'S SPRING.

——— "the leafy arbute spreads
A snow of blossoms, and on every bough
Its vermeil fruitage glitters to the sun."

ELTON.

This tree is a native of Greece, Palestine, and many other parts of Asia; of Ireland, and of many parts of the south of Europe. In Spain and Italy the country-people eat the fruit, which is said to have been a common article of food in the early ages. Virgil recommends the young twigs for goats in winter:

> ——— "Jubeo frondentia capris
> Arbuta sufficere."

It was used in basket-work:

> "Arbuteæ crates, et mystica vannus Iacchi."

Arbutus and oak formed the bier of the young Pallas, the son of Evander:

> "Haud segnes alii crates et molle pheretrum
> Arbuteis texunt virgis et vimine querno,
> Extructosque toros obtentu frondis inumbrant."
>
> VIRGIL, ÆNEIS, lib. xi.

"Others, with forward zeal, weave hurdles, and a pliant bier of arbute rods, and oaken twigs, and with a covering of boughs shade the funeral bed high-raised."—DAVIDSON'S TRANSLATION.

Horace, too, speaks of it, and celebrates its shade:

> "Nunc viridi membra sub arbuto
> Stratus."

Millar, after giving some of these quotations, adds, "I hope we shall no more have the classical ear wounded by pronouncing the second syllable of Arbutus long, instead of the first." This little ebullition of impatience, natural enough to a person who knew the right pronunciation, would have pleased his friend Dr. Johnson, who speaks of him somewhere as "Millar, the great gardener."

Some species of the Arbutus, from being mere shrubs, are better adapted for the present purpose than the beautiful one called the Common Strawberry-tree, which is the best known in our gardens; as the Painted-leaved, the Dwarf, and the Acadian Arbutus. These trees mostly like a moist soil, but the Acadian prefers a wet one: it is a native of swampy land, and if grown in a pot should be kept very wet: the earth, also, should be covered with moss, the better to retain the moisture. The other species should be watered every evening when the weather is dry, but not so liberally. When the frosts are severe, it will be more secure to shelter them; for though they will bear our winters when in the open ground, they are somewhat less hardy in pots. In mild seasons, a little straw over the earth would be a protection sufficient.

The berries of the Thyme-leaved Arbutus, which is a native of North America, are carried to market in Philadelphia, and sold for tarts, &c. Great quantities of them are preserved, and sent to the West Indies and to Europe. The London pastry-cooks frequently use these instead of cranberries, to which they are very similar; but they are inferior to cranberries of our own growth.

In Tuscany, many years ago, a man gave out that he had discovered a mode of making wine from the Arbutus. His wine was very good; but, upon his leaving the country, his wine-casks were found to contain a quantity of crushed grapes.

Upon the whole, the Arbutus, with its strawberry-like fruit, its waxen-tinted blossoms hanging in clusters, their vine-coloured stems, its leaves resembling the bay, and the handsome and luxuriant growth of its branches, is one of the most elegant pieces of underwood we possess: and when we have reason to believe that Horace was fond of lying under its shade, it completes its charms with the beauty of classical association.

ARUM.

CALLA ÆTHIOPICA.

AROIDEÆ. GYNANDRIA POLYANDRIA.

The Æthiopian species of this flower, commonly called the horn-flower, is the only one deserving of a place in the garden. Many Arums of the botanists are very useful as medicine, food, &c. and the leaves of the esculent Arum serve the inhabitants of the South-Sea islands for plates and dishes: but they are very little ornamental; and the few which are handsome have so powerful and disagreeable a scent as deservedly to banish them from most of our gardens.

This species, however, is exquisitely beautiful, and not only inoffensive in odour, but even agreeable. The leaves are large and glossy. It has a large white flower, folded with a careless elegance into the shape of a cup or bell, with a bright golden rod (called the spadix) in the centre. Placed by the side of the dark red peony, the effect is truly splendid: the contrast makes both doubly magnificent. A heathen might have supposed these fine flowers created on purpose to grace the bosom of the stately Juno. By the side of the rose, too, or the large double tulip, or some of the finer kinds of marygold, it has a noble appearance; and no flower is more deserving of care in the cultivation. In summer, the Arum should be allowed a liberal draught of water every evening; but, being a succulent plant, should be watered only at the roots. It flowers in May, and may stand abroad until the end of October: it should then be housed, and, during the winter, should be watered but once a week. It retains its leaves all the year: new ones displacing the old as they decay. In August the root should be taken out of the earth, when there will probably be a number of offsets upon it: these must be taken off, and planted in separate pots. The mother plant must then be carefully re-set in fresh earth, and, as well as the young roots, be placed in the shade until they have fixed themselves. In winter, although housed, it should be allowed plenty of fresh air in mild weather, and towards the end of April may be gradually accustomed to the open air.

The true Arums are similar plants, which, in a wild and humble state, are well known to children under the appellation of lords and ladies. Their natural stateliness gets them a fine name, in spite of their situation *.

ASPHODEL.

ASPHODELUS.

ASPHODELEÆ. HEXANDRIA MONOGYNIA.

King's-spear.—*French,* asphodéle.—*Italian,* asfodelo.

THE yellow Asphodel † is a native of Sicily, flowering in May and June: the white species ‡, a native of the south of Europe, flowers in June. The Onion-leaved Asphodel is a native of France, Spain, and the island of Crete: it flowers from June to August. The two last bear a starry flower, streaked with purple.

* They are also called Wake Robin; cuckow pint; ramp. In French, le gouet commun; bonnet de grand prêtre [high-priest's mitre]; herbe a petre; cheval bayard [bay horse]; pain de lievre [hares' bread].

† In French, la verge de Jacob [Jacob's staff].

‡ In French, hache royale, bâton royal, both signifying the royal sceptre.—In Italian, cibo regio [royal food].

They are tolerably hardy, the white least so; but they will all bear the open air, except in severe frosts, from which they require some protection. In dry summer-weather they should be watered every evening; in winter, once a week will suffice. The last-mentioned kind is an annual, and decays toward the end of October. It should be sown in the autumn: one seed in a pot. The first two species, as they do not flower the first year, will be better raised in a nursery: the first, when once obtained, may be increased by parting the roots, which should be done after the flower decays. They should be planted about two inches deep in the earth.

Rapin, in his poem on gardens, speaks of the Asphodel as an article of food:

> "And rising Asphodel forsakes her bed,
> On whose sweet root our rustic fathers fed."
>
> GARDINER'S TRANSLATION.

It is mentioned by Milton as forming part of the nuptial couch of Adam and Eve in Paradise:

> ——— "flowers were the couch,
> Pansies, and violets, and asphodel,
> And hyacinth, earth's freshest, softest lap."

It was formerly the custom to plant Asphodel and mallow around the tombs of the deceased. St. Pierre, after dwelling with some earnestness on the propriety of such customs, quotes the following inscription, engraven on an ancient tomb:

> "Au-dehors je suis entouré de mauve et d'asphodele, et au-dedans je ne suïs qu'un cadavre."

The fine flowers of the Asphodel produce grains, which, according to the belief of the ancients, afforded nourishment to the dead. Homer tells us, that having crossed the Styx, the shades passed over a long plain of Asphodel*. Orpheus, in Pope's Ode on St. Cecilia's day, conjures the infernal deities—

> " By the streams that ever flow,
> By the fragrant winds that blow
> O'er the Elysian flowers;
> By those happy souls who dwell
> In yellow meads of asphodel,
> Or amaranthine bowers."

Pope, according to a passage in Spence's Anecdotes, where he speaks of it with a disrespect hardly becoming a poet, seems to have thought it one of our commonest field-flowers.

ASTER.

CORYMBIFERÆ. SYNGENESIA POLYGAMIA SUPERFLUA.

Starwort, so named from its starry shape.—*French,* astére.—*Italian,* astero.

THE varieties of the Aster are infinite; and being very showy, of almost every colour, and the colours remarkably vivid, they make a brilliant figure in our gardens in the autumn. The most general favourite is the Chinese, or China Aster, which has larger and handsomer flowers than any of the others. There are many varieties of this species; white, blue, purple, and red; single and double of each; and another variety, variegated with blue and white.

The French call the China Aster *la Reine Marguerite,* which has been rendered, in English, the Queen Mar-

* See St. Pierre's Harmonies de la Nature.

garet: may they not rather mean to call it the Queen Daisy—*marguerite* being their name for the daisy, which this flower much resembles in form, though it is of a much larger size, and of more brilliant colours?

The Amellus, or Italian Starwort*, has a large blue and yellow flower. The leaves and stalks being rough and bitter, are not eaten by cattle; and thus remaining in the pastures after the grass has been eaten away, it makes a fine show when in full flower. This is supposed to be the Amellus of Virgil:

> "The Attic star, so named in Grecian use,
> But call'd amellus by the Mantuan muse."
>
> GARDINER'S TRANSLATION OF RAPIN.

> "Est etiam flos in pratis, cui nomen amello
> Fecere agricolæ; facilis quærentibus herba;
> Namque uno ingentem tollit de cespite silvam,
> Aureus ipse; sed in foliis, quæ plurima circum
> Funduntur, violæ sublucet purpura nigræ.
> Sæpe Deûm nexis ornatæ torquibus aræ.
> Asper in ore sapor: tonsis in vallibus illum
> Pastores, et curva legunt prope flumina mellæ.
> Hujus odorato radices incoque baccho;
> Pabulaque in foribus plenis appone canistris."
>
> VIRGIL, GEORGIC 4.

"We also have a flower in the meadows which the country-people call amellus. The herb is very easy to be found; for the root, which consists of a great bunch of fibres, sends forth a vast number of stalks. The flower itself is of a golden colour, surrounded with a great number of leaves, which are purple, like violets. The altars of the gods are often adorned with wreaths of these flowers. It has a bitterish taste. The shepherds gather it in the open valleys, and near the winding stream of the river Mella. Boil the roots of this herb in the best flavored wine; and place baskets full of them before the door of the hive."—MARTYN'S TRANSLATION, p. 390.

* Called in France l'œil de Christ [Christ's eye]; in Italy, amello, or astero affico di fior turchino.

The China Aster is an annual plant. It should be sown in March or April, and kept in a tolerably warm room until it has risen about three inches above the earth; and should then be gradually accustomed to the open air. The seed may either be sown singly, or many together, and removed into separate pots when they have grown about three inches: in the latter case, they must be placed in the shade until they have taken new root, and be gently watered every evening. According to their situation, China Asters will require water every evening, or second evening, in dry summer weather, after they are rooted; but it is necessary to give particular attention to this when they are newly planted. They will flower in August.

Most of the Asters have perennial roots and annual stalks, and may be increased by parting the roots, which should be done soon after the plant has done flowering. The Italian Starwort should not be removed oftener than every third year. The earth should be kept tolerably moist for all of them, and the taller kinds should be supported with sticks.

The African species must be raised in a hot-bed, and require protection in winter.

AUCUBA JAPONICA.

RHAMNEÆ? TETRANDRIA MONOGYNIA.

THIS tree, the leaves of which are singularly dabbled with spots, is very commonly grown in pots, as an ornament for balconies, windows, &c. and seems to have been long a favourite; probably, in some measure, from being of a hardy constitution, always green, and requiring little care—for it is by no means so handsome as many which are less generally regarded. It will bear the open air all the year round: the earth should be kept tolerably moist. Some call it American Laurel.

AURICULA.

PRIMULA AURICULA.

PRIMULACEÆ. PENTANDRIA MONOGYNIA.

Mountain Cowslip, French Cowslip, and Oricolo; but all these names have been superseded by Auricula, by which name it is best known in this country. The old botanical name was auricula ursi [bear's ear], from the shape of the leaves.—*French,* oreille d'ours.—*Italian,* orecchio d'orso.

THE Auricula is a native of the mountains of Switzerland, Austria, Styria, Carniola, Savoy, and Piedmont. It flowers in April and May. It is astonishing how greatly it may be improved by cultivation. It has been affirmed that Henry Stow, of Lexden, near Colchester, a noted cultivator of these flowers, had one plant with no less than one hundred and thirty-three blossoms upon one stem*.

The varieties are innumerable; and they are known by the name of every colour, and combination of colours. Some are named from the persons who first raised them; others by more fanciful appellations, as the Matron, the Alderman, the Fair Virgin, the Mercury, &c.

A fine Auricula should have a strong upright stem, of such a height that the flowers may be above the foliage of the plant. The foot stalks should also be strong, and

* Morant's Colchester (to which Millar refers), page 92.

proportioned in length to the size and number of the flowers, which should not be less than seven. The tube, eye, and border should be well-proportioned; that is, the diameter of the tube one-sixth, and that of the eye (including the tube) one-half the diameter of the whole flower. The circumference of the border should be a perfect circle; the anthers should be large, and fill the tube; and the tube should terminate rather above the eye, which should be very white, smooth, round, and distinct from the ground-colour. The ground-colour should be bold, rich, and regular, whether in a circle, or in bright patches: it should be distinct at the eye, and only broken at the outer part into the edging. The dark grounds are usually covered with a white powder, which seems necessary to guard the flower from the scorching heat of the sun.

Perhaps there is no flower more tenderly cherished by the cultivators than the Auricula: they wait upon and watch over it like a mother over her infant.

——————" Auriculas, enrich'd
With shining meal o'er all their velvet leaves."
THOMSON.

One Auriculist (for the science deserves a separate appellation) has devoted a little volume to its culture. An aspirant in this science is apt, however, to be startled on learning that the object of his adoration has a singular propensity for meat, and that a good part of its bloom is actually owing, like an alderman's, to this consumption of flesh. Juicy pieces of meat are placed about the root, so that it may in some measure be said to live on blood. This undoubtedly lessens its charms in some eyes. Its florid aspect somehow becomes unnatural; and the " shining meal," with which Thomson says it is " enriched," being no longer associated with vegetation, makes it look like a baker covered with flour, and just come out from a dinner in his hot oven.

The Auricula does not flower the first year; but as it is sometimes desirable to continue the handsome kinds, it may be occasionally agreeable to sow the seeds at home: directions are therefore given for that purpose. The seeds may be sown any time before Christmas, but the best time is in August. They may at first be sown within an inch of each other, not more than a quarter of an inch deep. They should stand in a moderately warm room, and be kept tolerably moist, by sprinkling the earth with a hard clothes-brush dipped in water, warmed by standing in the sun. At the end of four or five weeks, when the plants are all come up, they must be gradually accustomed to the air. As soon as any of the plants show six leaves, transplant them into other pots, about two inches asunder; and, when grown so as to touch each other, transplant them again, separately, into small pots, where they may remain to blow; and place them where they may enjoy the morning sun. Towards the middle of March they should be placed where they may receive the early, but be screened from the noon-day sun. Exposure to a whole day's sun at this time will destroy them; but, if the weather be mild, fresh air may be admitted to them. About the end of April they should be gradually accustomed to the open air; but care must be taken not to do this too abruptly, and to place them out on a mild day,

> "When dews, heaven's secret milk, in unseen showers,
> First feed the early childhood of the year."
>
> DAVENANT.

Special care must be taken to screen them from easterly winds. Earth, properly prepared for Auriculas, may be obtained from a nursery; and this is considered of some importance. What further directions are necessary will equally apply to those flowers raised at home, and to such as are only adopted children.

Preserve the plants from too much wet in winter, but let them have as much air as possible. To screen them from rain, it is best to keep them under cover. In February, when the weather is mild, take out of the pots as much of the earth as you can without disturbing the roots, and fill them up with fresh earth, which will greatly strengthen the plants: also take off such leaves as are decayed.

Auriculas should, in dry weather, be very gently watered three times in a week, carefully observing that no water fall upon the flowers; which, by washing off their farina, would greatly deface their beauty, and hasten their decay.

The best situation for Auriculas, when in bloom, is where the air may surround them, but roofed over head at such a distance as not to oppress the plants. Placed in an eastern balcony, shaded by a veranda, and by a few shrubs on the southern side, they will be well lodged. When the flowers have lost their beauty, they must be entirely exposed, to perfect their seeds, which will ripen in June. When the seeds are ripe, the seed-vessel will turn brown, and open. When they are perfectly dry, gather them, and lay them in an open paper exposed to the sun. To prevent their growing mouldy, they must remain in the pods till the season for sowing them.

Soon after they are past flowering, Auriculas should be taken out of the earth, such fibres as have grown very long should be shortened, and the lower part of the main root, if too long or decayed, cut off. If the lower leaves be faded or withered, strip them off in a downward direction: take off the offsets, and plant them in pots. Have ready a pot, three-parts filled with the prepared earth, highest in the middle; there place the old plant, with its fibres regularly distributed all round: then fill the pot up with the same earth, and lay a little clean coarse sand on the surface, round the stem of the plant. The pot should be gently shaken, to settle the earth about the root. It should be planted within half an inch of the lowest leaves; for, as the most valuable fibres shoot from that part, they will so be encouraged to strike root sooner.

When the offsets have formed one or more fibres of an inch or two in length, they may be parted from the mother-plant with the fingers, and planted as directed for young seedlings, several in a pot, until they are large enough to be transplanted separately.

In May, that is, as soon as this planting and transplanting is finished, the plants, old and young, should be placed in a shady, airy situation; by no means where the water from other plants can drip on them; and there remain till September, or, if the weather be mild, till October, when they must be sheltered from rain, snow, and frost, but must still be allowed air. They may be placed near a window, which should be open in mild weather, and closed when frosty.

Should there be offsets in April, or earlier, they may be taken off, and planted, without waiting till the old plants are removed. The following spring they will produce flowers, though but weakly. When past flowering, remove them into larger pots; and the second year they will flower in perfection. When the old plants are transplanted, they should, if requisite, be removed into larger pots.

It must be either the Auricula or the Polyanthus described by the poet in the following passage:

> "Oft have I brought thee flowers, on their stalks set
> Like vestal primroses, but dark velvet
> Edges them round, and they have golden pits."
>
> KEATS'S ENDYMION.

The Auricula is to be found in the highest perfection in the gardens of the manufacturing class, who bestow much time and attention upon this and a few other flowers, as the tulip and pink. A fine stage of these plants is scarcely ever to be seen in the gardens of the nobility and gentry, who depend upon the exertions of hired servants, and cannot therefore compete in these nicer operations of gardening with those who tend their flowers themselves, and watch over their progress with paternal solicitude.

AZALEA.

RHODORACEÆ. PENTANDRIA MONOGYNIA.

Azalea is derived from the Greek, and signifies dry.

MILLAR says the Azalea is so named because it grows in a dry soil; but this must be a strange oversight—for in the next page he tells us that it grows naturally in a moist soil, in North America, and that unless it has a moist soil it will not thrive.

The Azalea is a beautiful flowering shrub. The naked-flowered Azalea, in its native country, grows fourteen or fifteen feet high: here it is never more than half that height. Of this species, the flowers appear before the leaves: they are red, or white and red, and in great abundance. This shrub is common in the woods of New

Jersey, and is called May-flower, Wild Honeysuckle, and Upright Honeysuckle. We call it American Honeysuckle.

The White-flowered Azalea is a lower shrub than the former: the flowers are sweet-scented. This also is an American. The Pontic Azalea has yellow flowers. The Indian Azalea has a profusion of flowers, of a beautiful bright red.

The Azaleas should be sheltered from severe frost, and the earth be kept moist. They flower from May to July, and are too handsome to be dispensed with, but from absolute want of room.

BALM.

MELISSA.

LABIATÆ. DIDYNAMIA GYMNOSPERMIA.

From the fondness of bees for this plant, it is named melissa [a bee], melissophyllum [bee-leaf], from the Greek; and apiastrum, of a like signification, from the Latin. From its strong scent of lemons, Gesner has called it citrago.—*French,* le melisse des jardins [garden balm]; herbe de citron [lemon herb]; citronade, citronelle, both from the odour; poncirade; piment des mouches à miel [bees' spice].—*Italian,* melissa; cedronella; cedrancella; citraggine; melacitola.—In the Brescian territory, sitornela.

It is seldom that this darling of the bees is admitted into the flower-garden, yet it is very pretty when in flower; particularly that which is called the Great-flowered Balm, which has large purple flowers. Many a useless plant is admitted into the flower-garden with not half the beauty of this, which would deserve a place there for its scent alone. It was formerly considered as an efficacious remedy in hypochondria, but it is not so highly esteemed by the physicians of the present day. It proves, at least, an innocent substitute for foreign tea, which many persons find injurious to them; and many think its aromatic flavour very agreeable. Much of the prejudice against our native tea-plants has arisen from the tea being made of the fresh herbs, and by far too strong. If the Chinese tea were used as lavishly, it would be still more disagreeable to the taste than our native teas.

On account of its being so great a favourite with the bees, it was one of the herbs directed by the ancients to be rubbed on the hive, to render it agreeable to the swarm:

"Intorno del bel culto e chiuso campo
Lieta fiorisca l'odorata persa,
E l'appio verde, e l'umile serpillo,
Che con mille radici attorte e crespe
Sen va carpon vestando il terren d'erba,
E la melissa ch' odor sempre esala;
La mammola, l'origano, ed il timo,
Che natura creò per fare il mele."

LE API DEL RUCELLAI.

"O'er all the lawny field, lovely, shut in,
Let the glad violet smile with its sweet breath;
And parsley green; and humble creeping-thyme,
Which, with a thousand roots, curling and crisp,
Goes decking the green earth with drapery;
And balm that never ceases uttering sweets;
And hearts-ease, and wild marjoram, and thyme,
Which nature made on purpose to make honey."

"Quand' escon l'api dei rinchiusi alberghi,
E tu le vedi poi per l'aere puro
Natando in schiera andare verso le stelle,
Come una nube che si sparga al vento;
Contempla ben perch' elle cercan sempre
Posarsi al fresco sopra un verde elce,
Ovver presso a un muscoso e chiaro fonte.
E pero sparga quivi il buon sapore
De la trita melissa, o l'erba vile
De la cerinta; e con un ferro in mano
Percuoti il cavo rame, o forte suona
Il cembal risonante di Cibelle.

Questo subito allor vedrai posarsi
Nei luoghi medicati, e poi riporsi
Second il lor costume entr' a le celle."

LE API DEL RUCELLAI.

"When the bees issue from their nestling homes,
And you behold them through the clear blue ether,
Swimming tow'rd heaven like a wind-sprinkled cloud,
Be on the watch; for then it is they go
To feel the open air on a green oak,
Or near a mossy and fresh-bubbling fountain;
There follow them, and put the genial flavour
Of the bruised balm, or cerinth, and strike up
The hollow brass or tremble-touching cymbal,
And you will see them suddenly come down
Upon the season'd place, and so re-enter
After their wonted fashion, in their cells."

Virgil, in one of his pastorals, which was indeed the original of the poem of Rucellai, mentions green casia, wild thyme, and savory, instead of the violet, parsley, and wild thyme. By casia, some have supposed the poet intended rosemary; but in another passage he distinguishes these two plants: and as he uses the epithet 'green,' which the ancient poets almost invariably apply to parsley, it is probable Rucellai may have considered this as the plant described by Virgil. The frequent changes in the names of plants have occasioned much doubt and difficulty in ascertaining exactly the plants intended by old authors. Vaccinium has been translated by different writers, the privet, the hyacinth, the violet, &c.

Evelyn tells us that "this noble plant yields an incomparable wine;" and that "sprigs, fresh gathered, put into wine in the heat of summer, give it a marvellous quickness."

There is a plant called Bastard Balm, or Balm-leaved Archangel; in French, *Le Melissot*, or *Melisse de Punaisse* [Bug-balm]; of which the botanical name, Melittis, is similar in its etymology to Melissa. This, like the true Balm, yields a great deal of honey; it is described as having an unpleasant smell when fresh, but becoming delightfully fragrant when dried. It has large white and purple flowers, which are odoriferous when they first open. This plant is very handsome, and is a common inhabitant of the flower-garden.

Both these plants may be increased by parting the roots, which may be divided into pieces, with five or six buds to each, and planted in separate pots: this should be done in October. When intended for ornament, the roots should not be disturbed oftener than every third year. The earth should be loamy, and they should be placed in an eastern aspect, where they will thrive and produce flowers in abundance. The Melissa will flower in June or July; the Melittis, a month earlier. They may have a little water in dry weather, and stand abroad throughout the year. In autumn cut off the decayed stalks; new ones will grow in the spring.

BALSAM.

IMPATIENS.

BALSAMEÆ. SYNGENESIA MONOGAMIA.

Latin, impatiens.—*Italian,* balsamina; maraviglia di Francia; [the wonder of France].—In *Florence,* begl'uomini; bell' uomo [fine man].—*French,* balsamine, or belsamine. The Yellow Balsam is also called noli-me-tangere [touch me not]; quick-in-hand and wild mercury.—*French,* la balsamine des bois [Balsam of the woods]; la merveille; l'herbe Sainte Catharine; ne me touchez pas.—*Italian,* erba impaziente; balsamina gialla [Yellow Balsam].

Some of the names given to this plant refer to the violence with which the ripe seeds dart from the seed-vessel when touched.

In the day-time the leaves of this plant are expanded, but at night are pendent; contrary to the habit of plants in general, which are more apt to droop during the heat of the day. This plant grows in England and many other parts of Europe, and in Canada: it is the only species of Impatiens which grows wild in Europe.

The Garden Balsam, which, as its name implies, is the most commonly cultivated in our gardens, is a native of the East and West Indies, China, and Japan. The Japanese use the juice prepared with alum to dye their nails red. This beautiful flower has been much enlarged, and numerous varieties have been produced, by culture. Mr. Martyn, in his edition of Millar's Dictionary, speaks of having seen one, " the stem of which was seven inches in circumference, and all the parts large in proportion; branched from top to bottom, loaded with its party-coloured flowers, and thus forming a most beautiful bush."

There are white, purple, and red; striped and variegated, single and double, of each. Millar mentions two remarkable varieties:—the Immortal Eagle, a beautiful plant with an abundance of large double scarlet and white, or purple and white flowers;—and the Cockspur, of which the flowers are single, but as large as those of the former species; with red and white stripes. This is apt to grow to a considerable size before it flowers; so that in bad seasons it will bear but few blossoms.

In Ceylon and Cochin-China, there is a species of Balsam, from the leaves of which the inhabitants of Cochin-China make a decoction to wash and scent their hair.

The flowers of the Balsam will be handsomer if the plant be raised in a hot-bed: in May, if the weather be mild, it may be gradually accustomed to the open air. It must be watered every evening, but gently; and being a succulent plant, great care must be taken not to let water drip on it, nor to sprinkle it on the leaves or flowers. It loves the shade, and will thrive the better if shaded from the mid-day sun by the intervention of some light shrub, as the Persian lilac, &c. The Balsam is a general favourite for the number and beauty of the flowers, their sweetness, and the uprightness and transparency of its stem:

"Balsam, with its shaft of amber,"

says the poet, and the propriety of the expression has been questioned; but the introduction of a Balsam in the sunshine, not only fully justified its propriety, but excited surprise in those who had questioned it, at their own want of observation.

BASIL.

OCYMUM.

LABIATÆ. DIDYNAMIA GYMNOSPERMIA.

Basil is from a Greek word, signifying royal. It is generally called sweet basil.—*French,* basilic; la plante royale—*Italian,* basilico; ozzimo.—Ocymum is from a Greek word signifying *swift,* because the seed when sown comes up very quickly.

Basils are either herbs, or undershrubs, generally of a sweet and powerful scent: they are chiefly natives of the East Indies, and in this climate require protection from frost. They are raised in a hot-bed, but should have as much air as possible in mild weather. They may stand abroad from May to the end of September, or of October, according as the weather is more or less mild at this season. They should be kept moderately moist.

Many of the Basils will not live in this country, unless in a hot-house, but there are many that will, and among those are some of the handsomest and sweetest kinds; as the American Basil, with a flesh-coloured flower, remarkable for its agreeable scent; the Monk's Basil, a small annual plant, with a white and purple flower,—a mysterious foreigner, whose country is unknown to us; and Sweet Basil, which has spikes of white flowers, five or six inches in length, and a strong scent of cloves: of this species there is a variety smelling of citron, and another of which the flowers are purple.

In the East this plant is used both in cookery and medicine, and the seeds are considered efficacious against the poison of serpents.

The Basil, called by the Hindoos, holy or sacred herb, is so highly venerated by them, that they have given one of its names to a sacred grove of their Parnassus, on the banks of the Yamuna.

In Persia (where it is called rayhan), it is generally found in churchyards:

> ———— "the Basil-tuft that waves
> Its fragrant blossom over graves."

It is probably the custom to use it in Italy also to adorn tombs and graves, and this may have been Boccaccio's reason for selecting it to shade the melancholy treasure of Isabella. The exquisite story which he has told us, has lately become familiar to English readers, in the poems of Mr. Barry Cornwall and Mr. Keats. The former does not venture, like Boccaccio, to describe Isabella as cherishing the head of her lover, but makes her bury the heart in a pot of Basil; first so enwrapping and embalming it as to preserve it from decay. Mr. Keats is more true to his Italian original, and not only describes her as burying the head, but makes the head itself serve to enrich the soil, and beautify the tree; nay, even to become a part of it:

"And she forgot the stars, the moon, and sun,
And she forgot the blue above the trees,
And she forgot the dells where waters run,
And she forgot the chilly autumn breeze:
She had no knowledge when the day was done,
And the new morn she saw not,—but in peace
Hung over her sweet basil evermore,
And moisten'd it with tears unto the core.

"And so she ever fed it with thin tears,
Whence thick and green and beautiful it grew,
So that it smelt more balmy than its peers
Of basil-tufts in Florence; for it drew
Nurture besides, and life from human fears,
From the fast mouldering head there shut from view;
So that the jewel safely casketed
Came forth, and in perfumed leafits spread."

This young poet now lies in an Italian grave, which is said to be adorned with a variety of flowers. Among them Sweet Basil should not be forgotten.

And here we are naturally led to the Bay-tree.

BAY.

LAURUS NOBILIS.

LAURINEÆ. ENNEANDRIA MONOGYNIA.

Greek, Daphne.—*Italian*, alloro; lauro.—*French*, laurier.

This Bay, by way of distinction, called the Sweet Bay, well justifies the epithet: the exquisite fragrance of the Bay-leaf, especially when crushed, is known to every one; even in our climate, where it ranks but as a shrub, and doubtless, in its native soil, where it grows to a height of twenty or thirty feet, the perfume would be still finer.

How many grand and delightful images does the very name of this tree awaken in our minds! The warrior thinks of the victorious general returning in triumph to his country, amid the shouts of an assembled populace; the prince, of imperial Cæsar; the poet and the man of taste, see Petrarch crowned in the Capitol. Women, who are enthusiastic admirers of genius in any shape, think of all these by turns, and almost wonder how Daphne could have had the heart to run so fast from that most godlike of all heathen gods, Apollo.

It is said, that turning a deaf ear to the eloquent pleadings of the enamoured god, she fled, to escape his continued importunities: he pursued, and Daphne, fearful of being caught, entreated the assistance of the gods, who changed her into a laurel. Apollo crowned his head with its leaves, and commanded that the tree should be ever after held sacred to his divinity. Thus it is the true inheritance of the poet; but when bestowed upon the conqueror, is only to be considered as an acknowledgment that he deserves immortality from Apollo's children.

Spenser, indignant at the slight shown to his illustrious father, speaks in a vindictive strain of the fair Daphne:

> "Proud Daphne, scorning Phœbus' lovely fire,
> On the Thessalian shore from him did flee;
> For which the gods, in their revengeful ire,
> Did her transform into a laurel-tree."
>
> SPENSER'S SONNETS.

This noble tree has often been confounded with the common laurel, which is of quite a different genus, bearing the botanical name of prunus laurocerasus. The Bay was formerly called Laurel, and the fruit only named Bayes; this has probably occasioned the mistake. The word Bay, indeed, is probably derived from Bacca, the name of the berry.

Thomson, as if resolved to have the right laurel at any rate, makes use of both:

——— "from her majestic brow
She tore the laurel, and she tore the bay."
THOMSON'S BRITANNIA.

The Bay not only served to grace triumphant brows, mortal and immortal, but was also placed over the houses of sick persons, from some superstitious notion of its efficacy. It adorned the gates of the Cæsars and high pontiffs. It was worn by the priestess of Delphi, who chewed some of the leaves and threw them on the sacred fire. Letters and dispatches sent from a victorious general to the senate, were wrapped in Bay-leaves; the spears, tents, ships, &c. were all dressed up with them; and, in the triumph, every common soldier carried a branch in his hand.

The Bay was in great esteem with the physicians, who considered it as a panacea. The statue of Esculapius, though perhaps with an allusion also to his father Apollo (who was the god of physic in general, as his son seems to have been of its practitioners), was adorned with its leaves. From the custom that prevailed in some places of crowning the young doctors in physic with this Laurel in berry (Bacca-lauri), the students were called Baccalaureats, Bay-laureats, or Bachelors. The term has, with some propriety, been extended to single men, as the male and female berries do not grow on the same plant; and it seems we might with equal correctness bestow the name upon unmarried ladies.

The decay of the Bay-tree was formerly considered by the superstitious as an omen of disaster. It is said that before the death of Nero, though in a very mild winter, all these trees withered to the root, (yet surely his death was no serious disaster!) and that a great pestilence in Padua was preceded by the same phenomenon. The Laurel had so great a reputation for clearing the air and resisting contagion, that during a raging pestilence Claudius was advised by his physicians to remove his court to Laurentium on that account. It was also supposed to resist lightning, of which Tiberius was very fearful, and it is said, that to avoid it he would creep under his bed, and shade his head with the boughs.

Mr. L. Hunt alludes to this power in the Bay, in his Descent of Liberty:

"Long have you my laurels worn,
And though some under leaves be torn
Here and there, yet what remains
Still its pointed green retains,
And still an easy shade supplies
To your calm-kept watchful eyes.
Only would you keep it brightening,
And its power to shake the lightning,
Harmless down its glossy ears,
Suffer not so many years
To try what they can bend and spoil,
But oftener in its native soil
Let the returning slip renew
Its upward sap and equal hue;
And wear it then with glory shaded,
Till the spent earth itself be faded."

W. Browne tells us also, that "Baies being the materials of poets' ghirlands, are supposed not subject to any hurt of Jupiter's thunderbolts, as other trees are."—(See note to page 8, vol. i.).

"Where bayes still grow (by thunder not struck down),
The victor's garland and the poet's crown."

(See W. Browne's Poems, vol. iii.)

It is remarkable that this beautiful tree, which is hardy, handsome, sweet, and an evergreen, to say nothing of classical associations, is so seldom and so sparingly cultivated in this country. Evelyn tells us "that some Bay-trees were sent from Flanders with stems so even and upright, and with heads so round, full, and flourishing, that one of them sold for twenty pounds; and, doubtless," adds he, "as good might be raised here, were our gardeners as industrious to cultivate and shape them. I wonder we plant not whole groves of them, and abroad, they being hardy enough, grow upright, and would make a noble Daphneon."

Virgil celebrates the filial affection of the Bay, where, speaking of the different methods of propagating trees, he says,

"Others have a thick wood arousing from their roots; as cherries, and elms: the little Parnassian bay also shelters itself under the great shade of its mother."—MARTYN'S TRANSLATION, p. 114.

This would not, perhaps, convey to us so strong a meaning, did we not know, as Evelyn informs us, that while young, this tree thrives not well any-where *but* under its "mother's shade; where nothing else will thrive."

The Bay is a native of Asia, and the southern parts of Europe: it is not uncommon in the woods and hedges in Italy. The Abbé St. Pierre observes, "that it grows in abundance on the banks of the river Peneus, in Thessaly, which might well give occasion to the fable of the metamorphosis of Daphne, the daughter of that river."

It may be raised from berries, suckers, cuttings, or layers: it will bear the open air, and, when grown to a tolerable size, requires no other care than to water it occasionally in dry weather, to prune it in the spring, and to shift it into a larger pot when it has outgrown the old one. In doing this, the earth must not be cleared from the roots. A Bay-tree must not be hastily dismissed when it appears dead, but should be preserved till the second year; for when past hope of recovery, they will often revive, and flourish again as well as ever.

The Bay, which is the meed of the poet, a poet only

can celebrate; and what flower or tree has been more highly celebrated than this tree, which the resemblance of its name to that of his mistress induced Petrarch to make the continual subject of his pen? Thus, in speaking of the commencement of his passion, he uses this figure:

> "Amor fra l' erbe una leggiadra rete
> D'oro e di perle tese sott' un ramo
> Del l' arbor sempre verde, ch' i tant' amo
> Benchè n' abbia ombre più triste, che liete:"
>
> SONNET 148.

> Love mid the grass laid forth a lovely net
> Of woven pearls and gold, under the veil
> Of that fair evergreen I love so well,
> Although its shade is sad to me while sweet.

Again:

> "Arbor vittoriosa e trionfale,
> Onor d' imperadori e di poeti,
> Quanti m' hai fatto di dogliosi e lieti
> In questa breve mia vita mortale!"
>
> SONNET 225.

> O thou victorious and triumphant tree,
> Glory of poets and of emperors,
> How many sad and how many sweet hours
> Hast thou in this short life bestow'd on me!

> "L' aura celeste; che 'n quel verde Lauro
> Spira, ov' Amor ferì nel fianco Apollo
> E a me pose un dolce giogo al collo
> Tal, che mia libertà tardi ristauro."
>
> SONNET 164.

> "L' aura che 'l verde lauro, e l' aureo crine
> Soavemente sospirando move;
> Fa con sue viste leggiadrette, e nove
> L' anime da' lor corpi pellegrine*.
>
> SONNET 208.

* The play upon the word Laura in these passages does not (as the Italian reader will readily perceive) easily admit of translation.

After the death of Laura, he writes:

> "Rotta è l' alta Colonna, e 'l verde Lauro,
> Che facean ombra al mio stanco pensero:"
>
> SONNET 229.

evidently alluding to the death of his mistress, and that of Cardinal Colonna; and a high compliment, indeed, it was to the cardinal, on such a subject to unite his name with hers.

How tender and how natural is the following sonnet:

> "Quand' io veggio dal ciel scender l' aurora
> Con la fronte di rose, e co' crin d' oro;
> Amor m' assale: ond' io mi discoloro;
> E dico sospirando, ivi è Laura ora.
> O felice Titon tu sai ben l' ora
> Da ricovrare il tuo caro tesoro:
> Ma io che debbo far del dolce Alloro;
> Che se 'l vo' riveder, conven ch' io mora.
> I vostri dipartir non son si duri;
> Ch' almen di notte suol tornar colei
> Che' non ha a schifo le tue bianche chiome:
> Le mie notti fa triste, e i giorni oscuri
> Quella, che n' ha portato i pensier miei;
> Nè di se m' ha lasciato altro, che 'l nome."
>
> SONNET 250.

Again I have to lament that the absence of a poetical friend will not allow me to add a proper translation of this sonnet. To give the English reader some notion of the *subject*, I have translated it in humble prose. I need not add, that this can convey but a very inadequate idea of the original:

> "When I behold Aurora descending from heaven, with her cheek of roses, and her locks of gold, love assails me: I turn pale, and I say, sighing, where is Laura now? Oh, happy Tithonus, thou knowest well the hour when thou wilt recover thy dear treasure: but what shall I do for the sweet laurel, which would I see again, I first must die! Your parting is less cruel; for night at least restores to thee her who scorns not thy white locks: she makes my nights sorrowful, and my days dark, who has borne away my thoughts, and of herself has left me nothing but the name."

But unless Petrarch's whole works are inserted, it will be a vain attempt to give all the passages in which he thus celebrates both his mistress and the tree. One or two more only shall be mentioned: the canzone beginning

> "Standomi un giorno solo a la fenestra;"
>
> CANZONE 42.

and

> "Quando il soave mio fido conforto."
>
> CANZONE 47.

It was but just that he should be crowned with this beloved Laurel, as it is well known that he was, publicly, at Rome; having been offered the same honourable distinction at Paris also.

"The Laurel seems more appropriated to Petrarch, (says Mr. Hunt,) than to any other poet. He delighted to sit under its leaves; he loved it both for itself and for the resemblance of its name to that of his mistress; he wrote of it continually; and he was called from out of its shade to be crowned with it in the Capitol. It is a remarkable instance of the fondness with which he cherished the united ideas of Laura and the Laurel, that he confesses it to have been one of the greatest delights he experienced in receiving the crown upon his head *."

Chaucer bestows the Laurel upon the Knights of the Round Table, the Paladines of Charlemagne, and some of the Knights of the Garter,

> "That in their timis did right worthily.
>
> * * * * *
>
> For one lefe givin of that noble tre
> To any wight that hath done worthily
> (An it be done so as it ought to be)
> Is more honour than any thing erthly,
> Witness of Rome; that foundir was truly
> Of all knighthode and dedis marvelous,
> Record I take of Titus Livius."

* Indicator, No. XL. vol. i. page 316.

Chaucer evidently intends the genuine Laurel, not the usurper of the title, since he speaks of its sweet scent:

"And at the last I gan full well aspy
Where she sate in a fresh grene laury tre,
On the furthir side evin right by me,
That gave so passing a delicious smell,
According to the eglantere full well."

THE FLOURE AND THE LEAFE.

The following lines, addressed by Tasso to a Laurel in his lady's hair, are, with their translation, taken from the Literary Pocket-Book for the year 1821:

"O pianta trionfale,
Onor d' imperatori,
Hor de' nomi de' regi anco t' onori
Cosi di pregio in pregio,
Di vittoria in vittoria,
Vai trapassando, e d' una in altra gloria;
Arbore gentile, e regio,
Per che nulla ti manchi, orna le chiome
Di chi d' Amor trionfa, e l' alme ha dome."

O glad triumphal bough,
That now adornest conquering chiefs, and now
Clippest the brows of over-ruling kings:
From victory to victory
Thus climbing on, through all the heights of story,
From worth to worth, and glory unto glory;
To finish all, O gentle and royal tree,
Thou reignest now upon that flourishing head,
At whose triumphant eyes Love and our souls are led.

BELVEDERE.

CHENOPODIUM SCOPARIA.

ATRIPLICEÆ. PENTANDRIA DIGYNIA.

Called also Summer Cypress.—*French*, la belvedère; bellevedere; belle a voir.—*Italian*, il belvedere: all which foreign names refer to its beautiful appearance.

THIS is an extremely handsome plant, growing very

close and thick, in the form of a pyramid, as regular as if cut by art: it has so much the appearance of a young cypress tree, that but for the leaves being of a more lively green, it might at a little distance be mistaken for one. It grows naturally in Carniola, Greece, China, and Japan.

The seeds should be sown in autumn, singly, or several together, and divided into separate pots in the spring, when they come up. In autumn, when they ripen their seeds, if other pots are standing pretty near, the seeds will be apt to fall into them, and the self-sown plants will come up the following spring: so that it will be well to keep such pots as will not admit of such an unceremonious visitor at a sufficient distance to secure them from intrusion. The earth should be kept moderately moist.

BITTER-VETCH.

OROBUS.

LEGUMINOSÆ. DIADELPHIA DECANDRIA.

French, l'orobe; pois de pigeon [pigeon's pea].—*Italian,* orobo; robiglia.

The Yellow Bitter-Vetch is described by Haller as one of the handsomest of the papilionaceous tribe. It is a native of Siberia, Switzerland, Italy, and the South of France. Spring Bitter-Vetch has a handsome flower, curiously shaded with red, purple, and blue, becoming altogether a sky-blue before it falls. It grows in the woods in many parts of Europe, and flowers in March and April. The Tuberous Bitter-Vetch, called also heath peas, wood peas, and in French *gesse sauvage,* has also a brilliant flower of red-purple, fading to a blue as it decays. The Highlanders, who call it corr, or cormeille, dry the tubercles of the root, and keep them in the mouth to flavour their liquor. They affirm, that they are enabled, by the use of them, to repel hunger and thirst for a long time. This idea reminds one of a passage in one of the Italian poets, where an enchanter preserves two knights from starvation during a long journey by giving them an herb, which, being held in the mouth, answers all the purposes of food.

The taste of these roots resembles that of liquorice-root, and, when boiled, they are well-flavoured and nutritive. In times of scarcity, they have served as a substitute for bread. The plant is a native of most parts of Europe. These, and the other hardy kinds, may be increased by parting the roots, which should be done in the autumn. They generally delight in shade, and prefer a loamy soil: the earth should be kept moderately moist.

BLOODWORT.

SANGUINARIA.

PAPAVERACEÆ. POLYANDRIA MONOGYNIA.

The English name is from its blood-coloured juice. It is also named, by the Americans, Puccoon.

"THOUGH the Sanguinaria cannot be considered as a showy plant," says Mr. Martyn, "yet it has few equals in point of delicacy and singularity: there is something in it to admire, from the time that its leaves emerge from the ground and embosom the infant blossom, to their full expansion, and the ripening of the seeds."

In the woods of Canada, and other parts of North America, it grows in abundance: the Indians are said to paint their faces with the juice. In this country the flowers open in April, but they fully expand only in fine warm weather.

We are told, that in the year 1680 this plant was cultivated in "Mr. Walker's suburban garden in St. James's

Street, near the palace." Its flowers are white, and three or four flower-stems spring from one root: it prefers a loose soil and a shady situation, and may be annually increased by parting the roots in September. When the flowers decay, the green leaves come out, which last till Midsummer: from which time till autumn the roots remain inactive. It should be planted in a pot seven or eight inches wide, and an equal mixture of bog earth and rotten leaves will be the best soil. It must be watered every evening in dry summer weather. The earth may be covered with moss, which will tend to preserve the moisture in the summer, and to protect the roots from frost in the winter.

BOX TREE.

BUXUS.

EUPHORBIACEÆ. MONOECIA TETRANDRIA.

French, le buis; le bois beni [blessed wood].—*Italian,* busso; bosso; bossolo; in the Brescian territory, martel [hammer wood]; buz.

PROPERLY speaking, there is but one species of Box; varying much in size, and somewhat in the colour of its leaves. It may be easily propagated both by seeds and cuttings; but is so slow of growth, as to be many years in attaining any considerable size. It is therefore advisable to purchase it of the size desired, rather than to raise it at home. It will thrive in any soil or exposure, and under the deepest shade. It is an evergreen, and remarkable for its fine glowing colour: particularly the dwarf kind. In the story of Rimini, it is called "sunny-coloured box." "The pleasantness of its verdure," says Evelyn, "is incomparable."

The Box-tree, though in gardens seldom seen more

than three or four feet high, will, if not cut, rise to a height of twelve or fifteen. The wood is close-grained, very hard, and heavy. It is the only one of the European woods that will sink in water; and is sold by weight, fetching a high price. Not being liable to warp, it is well adapted to a variety of nicer purposes; as tops, screws, chess-men, pegs for musical instruments, knife-handles, modelling-tools, &c. The ancients made combs of it, which use is mentioned by Cowley in his poem on Plants:

> "They tye the links that hold their gallants fast,
> And spread the nets to which fond lovers haste."

Corsican honey was supposed by the ancients to owe its ill name to the bees feeding upon Box: none of our animals will touch it. Parkinson says, "the leaves and saw-dust boiled in lye will change the hair to an auburn colour."

When it was the fashion to clip and cut trees into the shapes of beasts, birds, &c. the Box was considered as second only to the yew for that purpose; for which, Pliny says that nothing is better adapted. Martial notices this quality in speaking of Bassus's garden:

> ——— "otiosis ordinata myrtetis,
> Viduaque platano tonsilique buxeto."

> "There likewise mote be seen on every side
> The yew obedient to the planter's will,
> And shapely box, of all their branching pride
> Ungently shorne, and with preposterous skill,
> To various beasts, and birds of sundry quill
> Transform'd, and human shapes of monstrous size;
>
> * * * * *
>
> "Also other wonders of the sportive shears
> Fair Nature mis-adorning, there were found
> Globes, spiral columns, pyramids and piers
> With sprouting urns, and budding statues crown'd;

> And horizontal dials on the ground
> In living box by cunning artists traced;
> And gallies trim, on no long voyage bound,
> But by their roots there ever anchor'd fast,
> All were their bellying sails outspread to every blast."
>
> G. WEST.

This preposterous taste in gardening was at last reformed by the pure and classical taste of Bacon; who, though no enemy to sculpture, did not approve of this absurd species of it: at once disfiguring art and nature.

"In several parts of the North of England, when a funeral takes place, a basin full of sprigs of Box-wood is placed at the door of the house from which the coffin is taken up; and each person who attends the funeral, ordinarily takes a sprig of this Box-wood, and throws it into the grave of the deceased."—(See Note in WORDSWORTH'S POEMS, 8vo. vol. i. p. 163.)

> "The bason of box-wood, just six months before,
> Had stood on the table at Timothy's door;
> A coffin through Timothy's threshold had pass'd,
> One child did it bear, and that child was his last."
>
> WORDSWORTH.

Gerarde informs us, that turners and cutlers call Box-wood dudgeon, because they make dudgeon-hafted knives of it. The Box-tree is a native of most parts of Europe, from Britain southwards: it also abounds in many parts of Asia and America. In England it was formerly much more common than at present.

"These trees," says Evelyn, "grow naturally at Boxley in Kent, and at Box-hill in Surrey: giving name to them. He that in winter should behold some of our highest hills in Surrey, clad with whole woods of them, for divers miles in circuit, as in those delicious groves of them belonging to the late Sir Adam Brown of Beckworth Castle, might easily fancy himself transported into some new or enchanted country."

But this enchantment has been long since dissolved. Mr. Millar, in 1759, lamented the great havoc made among the trees on Box-hill, though there then remained several of considerable magnitude; but since that time the destruction has been yet greater. Not only this hill in Surrey, and Boxley in Kent, but Boxwell in Coteswold, Gloucestershire, is said to be named from the Box tree. It has been made a serious and heavy complaint against Box, that it emits an exceedingly unpleasant odour, of which the poets speak as a thing notorious: yet it is only when fresh cut that the scent is unpleasant, and a little water poured over it immediately removes this objection.

BROOM.

SPARTIUM.

LEGUMINOSÆ. DIADELPHIA DECANDRIA.

French, le genêt; le genêt a balais.—*Italian,* sparzio; scopa; ginestra: all referring to its use as besoms*.

The Brooms are very ornamental shrubs, with few leaves, but an abundance of brilliant and elegant flowers: they strike a deep root, but are too handsome to be rejected where room can be afforded for them. They must be planted in a pot or tub of considerable depth. There are three species with white, and one with violet-coloured flowers: the others have all yellow blossoms.

The violet-coloured has no leaves, and is usually called the Leafless Broom: it was found by Pallas in the Wolga Desert. The Spanish Broom has yellow—the Portugal, white blossoms. The white-flowered, one-seeded kind, is

* The family of Plantagenet took their name from this shrub, which they wore as their device.

a native both of Spain and Portugal. "It converts the most barren spot into a fine odoriferous garden," says Mr. Martyn, speaking of this species.

All the species here named will endure the cold without shelter: they do not like much wet. Our common Broom surpasses many of the foreign kinds in beauty: indeed, few shrubs are more magnificent than this evergreen, with its profusion of bright golden blossoms. They are the delight of the bees: and the young buds, while yet green, are pickled like capers. It is said that the branches are of service in tanning leather, and that a kind of coarse cloth is manufactured from them. The young shoots are mixed with hops in brewing: and the old wood is valuable to the cabinet-maker. Brooms are made from this shrub; and, from their name, it is supposed to have furnished the first that were made. In the north of Great Britain it is used for thatching cottages, corn, and hay-ricks, and making fences. In some parts of Scotland, where coals and wood are scarce, whole fields are sown with it for fuel.

But the Scotch have long been aware of the poetry as well as the utility of this beautiful shrub. The burden of one of their most popular songs is well known:

"O the broom, the bonny bonny broom,
The broom of the Cowden-knows;
For sure so soft, so sweet a bloom
Elsewhere there never grows."

Burns lauds it, too, in one of his songs, written to an Irish air, which was a great favourite with him, called the Humours of Glen:

"Their groves of sweet myrtle let foreign lands reckon,
Where bright beaming summers exalt the perfume;
Far dearer to me yon lone glen o' green breckan,
Wi' the burn stealing under the lang yellow broom.

"Far dearer to me are yon humble broom bowers,
Where the blue-bell and gowan lurk lowly unseen;
For there lightly tripping amang the sweet flowers,
A listening the linnet, oft wanders my Jean."

"'Twas that delightful season, when the broom
Full-flowered, and visible on every steep,
Along the copses runs in veins of gold."
WORDSWORTH'S POEMS, 8vo. vol. ii. p. 265.

Thomson speaks of it as a favourite food of kine. It flowers in May and June.

"Yellow and bright, as bullion unalloyed,
Her blossoms."
COWPER'S TASK.

Broom makes a pleasant shade for a lounger in the summer: it seems to embody the sunshine, while it intercepts its heat:

"To noontide shades incontinent he ran,
Where purls the brook with sleep-inviting sound;
Or, when Dan Sol to slope his wheels began,
Amid the broom he basked him on the ground,
Where the wild thyme and camomile are found."
CASTLE OF INDOLENCE, Canto 1.

Mr. Horace Smith speaks of it as poisonous, yet most of the species are eaten by cattle: some are particularly recommended as a food for kine. The Base Broom, or Green-weed, is said to embitter the milk of the cows that eat of it; but, from the bitterness of the plant itself, they commonly refuse it.

————————————" my herd
Cannot be browsed upon the mount, for so
The heifers might devour with eager tongue
The poisonous budding brooms."
AMARYNTHUS.

Browne alludes to the use of Broom in thatching:

> "Among the flags below, there stands his coate,
> A simple one, thatched o'er with reed and broom;
> It hath a kitchen, and a several room
> For each of us."
>
> BRITANNIA'S PASTORALS.

A Russian poet speaks of the Broom as a tree:

> "See there upon the broom-tree's bough
> The young grey eagle flapping now."
>
> BOWRING'S RUSSIAN ANTHOLOGY.

The blossom of the Common Broom closely resembles that of the Furze, both in form and colour—that Furze which sheds such a lustre over our heaths and commons, and at sight of which, it is said, Dillenius fell into a perfect ecstacy. In many parts of Germany the Furze-bush is unknown. Gerarde says, that about Dantzic, Brunswick, and in Poland, there was not a sprig of either Furze or Broom; and it is really a striking sight to come suddenly upon a common, glowing, as it were, in one great sea of gold. Gerarde adds, that, in compliance with earnest and repeated entreaties, he sent seeds to these places, and that the plants raised from them were curiously kept in the finest gardens. Furze bears various names in different parts of England: Furze in the south, Whin in the east, and Gorse in the north.

> "The prickly gorse, that, shapeless and deformed,
> And dangerous to the touch, has yet its bloom,
> And decks itself with ornaments of gold."
>
> COWPER'S TASK.

> "Or from yon swelling downs, where sweet air stirs
> Blue harebells lightly, and where prickly furze
> Buds lavish gold."
>
> KEATS'S ENDYMION.

St. Pierre evidently alludes to the Furze-bush in the following passage: "I saw in Brittany a vast deal of uncultivated land; nothing grows upon it but Broom, and a shrub with yellow flowers, which appeared to me a composition of thorns. The country-people called it Lande, or San: they bruise it, to feed their cattle. The Broom serves only to heat their ovens. It might be turned to better account. The Romans made cord of it, which they preferred to hemp, for their shipping."—St. Pierre's Voyage to the Isle of France.

It is also called in different parts of France, *Jonc Marin* [Sea-rush]; *Porc Marin* [the Sea-hog]; *Lande Epineuse* [Thorny Heath]. Its botanical name is Ulex.

BROWALLIA.

BROWALLIEÆ. DIDYNAMIA ANGIOSPERMIA.

So named by Linnæus, from Job Browallius, Bishop of Aboa.

This is but an annual plant, and must be raised in a hot-bed; but it is worth procuring for its short-lived beauty, on account of the extreme brilliancy of the colours. "We cannot," says Mr. Curtis, "do justice to it by any colours we have." There are but two kinds: the Upright, and the Branching. The former is the handsomest. It is a native of Peru, and flowers from July to September. It should be kept within doors till June; and, in dry and hot weather, should be frequently, but sparingly, watered.

CAMELLIA JAPONICA.

CAMELLIADEÆ. MONADELPHIA POLYANDRIA.

So named in honour of Joseph Kamel, a jesuit, whose name is usually spelled Camellus. This tree is sometimes called Japan Rose.

THIS beautiful evergreen must be sheltered from the middle of September till the beginning of June. In the summer, when the weather is dry, it should be watered every evening, or second evening, according to the heat of the sun: in the winter once a week will suffice, and that should be at noon. There are double and single varieties; white, purple, and red of each. This tree has the appearance of a bay bearing roses, much more than the rhododendron, which, from some fancied resemblance of that sort, is also named rose-bay.

There are several other Camellias, requiring the same treatment as this, which is the handsomest species. Had the Camellia been a Greek, or Italian, or English plant, there would have been a great deal said of it by poets and lovers; and doubtless it makes a figure in the poetry of Japan. But, unfortunately for our quotations, though perhaps fortunately for their own comfort, the Japanese have hitherto had most of their good things to themselves. Their country would lay open a fine field for the botanist. See an interesting account of this apparently intelligent and amiable people in Golownin's Narrative of his Captivity among them.

There are two superb collections of the Camellia Japonica open to the public: one at Vauxhall, the other at Hackney.

CAMPANULA.

CAMPANULACEÆ. PENTANDRIA MONOGYNIA.

Italian, campanella.—*French,* campanule, or campanette.—*English,* Bell-flower. These names signify a little bell, and were given to the flower on account of its bell-like shape.

MILLAR mentions seventy-eight kinds of Campanula, of which it will be sufficient to specify some of the most desirable; as the Venus's Looking-glass, which has usually a handsome purple flower, but sometimes white. This plant takes its name from the glossiness of the seeds. It is also called Corn-Gilliflower, and Corn-Pink: in French, *Miroir de Venus,* but at Paris, *la Doucette:* in Italian, *Specchio di Venere.*

It is a native of the south of Europe. Plants sown in the autumn will flower in May, a month earlier than those sown in the spring. The seeds may be sown about an inch asunder; the earth should be kept moist, and the plant should remain in the open air. The roots of this species are annual.

The Peach-leaved Campanula is a perennial. The flowers are blue or white; double and single varieties of each. This may be increased by parting the roots, which should be done in September. It will thrive in any soil or situation.

The Giant Throatwort is a native of England and most parts of Europe. It has a purple or white flower, which blows in July and August. This species loves shade.

Great Throatwort, Canterbury Bells, called in French *la Cloche* [Bell], *la Clochette* [Little Bell], *les Gands de Notre Dame* [Our Lady's Gloves], is a native of Europe and Japan. It has purple or white flowers, blowing in July and August. This species may be increased in the same manner as the Peach-leaved, but prefers a loamy soil: they are both very hardy. The name of Throat-wort was given to these plants from a notion that they would cure inflammation and swelling of the throat.

The lesser Canterbury Bells have purple, brilliant blue, or white flowers, which continue from June to September. This prefers a dry chalky soil: in a rich soil the flowers are apt to lose their colour. This is the Calathian Violet; also called Autumn Bell-flower, Autumn Violet, and Harvest Bells.

The Medium, or Coventry Bells,—in French, *Mariettes*, and in Italian, *Viola Mariana* [Mary's Violet]—to which Gerarde gives the name of Mercury's Violets, have large and handsome flowers, blowing in June: their colours, blue, purple, white, or striped.

The Campanulas here enumerated, and such others as are not natives of the Cape, are sufficiently hardy to endure the open air in the winter, although some of them are sheltered while seedlings. Most of them may be increased by cuttings or seeds. Those raised from cuttings flower more quickly; those from seeds are considered as the strongest. They should be sparingly watered.

There is a species of Campanula which is trained to conceal fire-places in the summer, and has a very pretty effect when so used. It is the Pyramidal Campanula; *la Pyramidale des Jardins* of the French. The roots send out three or four strong upright stalks, which grow nearly four feet high, and are garnished with smooth oblong leaves and an abundance of large blue flowers. These upright stalks send out short side-branches, which are also adorned with flowers; so that, by spreading the upright stalks to a flat frame composed of slender laths, the whole plant is formed into the shape of a fan, and will perfectly screen a common sized fire-place. The plant may stand abroad till the flowers begin to open; and, being then placed in a room where it is shaded from the sun and rain, the flowers will continue long in beauty. If it be removed into the air at night, where it is not exposed to heavy rains, the flowers will be handsomer, and will last longer. This kind is rather more delicate than those before mentioned; and when raised from seeds, which is the best mode, requires a hot-bed to bring it forward. It should therefore be procured in a pot, and should be one that has been raised from seed. Most of the Campanulas close their flowers at night. They will grow in common garden earth.

CANDY-TUFT.

IBERIS.

CRUCIFERÆ. TETRADYNAMIA SILICULOSA.

Candy-Tuft takes its English name from Candia, one of the many countries of which it is a native; and its Latin name from Iberia, now Spain.

The evergreen kinds are more tender than most of the species, and require shelter from frost: they do not thrive so well in a pot as in the open ground, but cannot for a comparative inferiority be dispensed with. In addition to the advantage of retaining their green leaves all the year, they enliven the winter months with their tufts of white flowers, which continue in succession from the end of August till the beginning of June.

There are two species of evergreen Candy-Tufts: the broad and the narrow-leaved. The former is a native of Persia; the latter, of the island of Candia. As these do not often produce seeds in England, they are increased by cuttings, which may be planted in any of the summer months; and, if shaded from the sun, and kept moist, will take root in two months. Their branches will fall unless supported by sticks.

The Common Purple Candy-Tuft, the White, and the Sweet-scented are annuals; and, if sown in September, March, April, and May, may be continued in succession throughout the summer. These, as well as the Rock and the Round-leaved Candy-Tufts, will bear exposure to the open air. They must not have more water than is sufficient to keep them from absolute drought.

The Purple has a variety of names: as Candia Thlaspi, Candia Mustard, and Spanish Tuft. The White species, though not mentioned by any of the old botanical writers, is indigenous: it is common to most European countries. The Sweet-scented, the flowers of which are dazzlingly white, is a native of the mountains near Geneva. The seeds should be sown in pots four or five inches in diameter, one in each.

CARDAMINE.

CRUCIFERÆ. TETRADYNAMIA SILIQUOSA.

So called from its taste of cardamoms: also Lady's Smock, from the white sheets of flowers they display on the plashes of water in which they usually grow; and Cuckoo-flower, from blowing at the time of that bird beginning to sing.—*French,* cresson de pres [meadow-cress]; passarage sauvage [wild cress].—*Italian,* cardamindo; nasturzio di prato; o crescione di prato: both signifying meadow-cress.

Few of the species of Cardamine are admitted into gardens. The kind most deserving of a place there is the common Cuckoo-flower, or Lady's Smock, which is common in our meadows, and by brook sides, &c; or, rather, the double varieties of this kind should perhaps be selected. This flower has been usually described by the poets as of a silvery whiteness, which shows the season they have chosen for their rural walks to have been a late one; as, in its natural state, it is more or less tinged with purple, but becomes white as it fades, by exposure to the heat of the sun. "The allusions to the whiteness of the corollas," says Rousseau, " will not hold, for they are commonly purple."

The various shades of these flowers, with the little green leaves that enclose the unopened buds, have an exceedingly pretty effect when a quantity of them are collected; and if kept in fresh water, and well supplied, they will survive their gathering for a fortnight or more. The young leaves are eaten in salads.

The double varieties are white or purple: they are increased by parting the roots in autumn. They love the shade, and should be plentifully watered every evening. It is called the Cuckoo-flower, because it comes at the same time with the cuckoo; and, for the same reason, the name has been given to many other flowers. Shakspeare's Cuckoo-buds are yellow, and supposed to be a species of ranunculus. Indeed, he expressly distinguishes *his* Cuckoo-bud from this flower:

"When daisies pied, and violets blue,
And lady's-smocks all silver white,
And cuckoo-buds of yellow hue,
Do paint the meadows with delight."

" So have I seen a ladie-smock soe white,
Blown in the mornynge, and mowd down at night."

CHATTERTON'S BATTLE OF HASTINGS.

CARDINAL-FLOWER.

LOBELIA.

LOBELIACEÆ. SYNGENESIA MONOGAMIA.

Named from Matthias de Lobel, a Flemish botanist, physician and botanist to King James the First.—*French,* la cardinale.—*Italian,* fior cardinale; cardinalizia.

THE Cardinal-flower is a very handsome plant, the scarlet species in particular: the blue, however, is very handsome. They do not flower the first year: yet, as the offsets produced from the roots do not flower so strongly as seedling plants, it is better to sow them. This should be done in the autumn. They may at first be sown several together: the pots in which they are sown should stand abroad in mild weather, but under cover in frost or heavy rain. In spring the plants will appear. They may then remain abroad altogether, and must be kept always rather moist. When big enough to remove, they may be replanted separately into small pots; or, if preferred, may be so sown at first. They should be placed where they may enjoy the morning sun, and there remain till autumn: they must then be taken into the house, but stand near an open window in mild weather. If in the course of the summer the roots should fill the pots, the plants must be removed into larger ones. The following spring they must be potted in fresh earth, and again placed abroad. They will flower in August; and, if not exposed to the mid-day sun, will continue long in beauty. The roots will last two or three years. They are likewise increased by their offsets, and by cuttings of the stalks, like rockets; but no other way is so good as sowing them.

CATCHFLY.

SILENE.

CARYOPHYLLEÆ. DECANDRIA TRIGYNIA.

French, le cornillet; attrape mouche [catch fly.]

THIS plant is covered with a glutinous moisture, from which flies, happening to light upon it, cannot disengage themselves. This circumstance has obtained it the name of Catchfly; to which Gerarde adds the name of Limewort.

If the seeds are sown in the autumn, separately, in pots about six inches in diameter, and in a dry soil, they will grow without further attention. They will bear the open air; and, unless in very dry weather, will not need watering. These directions will serve for nearly all the kinds, of which there are upwards of sixty. There are, however, two exceptions: the Dark-flowered and the Waved-leaved species, which require a stove.

CELANDINE.

CHELIDONIUM.

PAPAVERACEÆ. POLYANDRIA MONOGYNIA.

The name of this plant is derived from the Greek, and signifies a swallow. It is not so named, as some have supposed, from its coming and going with the swallow; but, according to Gerarde, from an opinion which prevailed among the country-people, that the old swallows used it to restore sight to their young *when their eyes were out.* For the same reason it is also called Swallowwort.

THE Sea Celandine, or Yellow Horned Poppy (called also Bruisewort), is a flower common to every part of Europe, growing on sandy soils, chiefly by the sea-shore.

The flowers fall the second day after they are blown; but they are large, form a fine contrast with the sea-green colour of the leaves, and follow each other in such quick succession and abundance almost all the summer, as to make it a valuable plant. It begins to flower in June. It is a perennial flower. The whole plant abounds with a poisonous juice, which is said to occasion madness.

The Red and the Violet Celandines, or Horned Poppies, are common in Europe, growing in the same sandy soil as the former. These flower in July and August.

The Great, or Major Celandine, is common in hedges, and other shady places; on rubbish, rocks, or old walls*. It bears a bright yellow flower, and continues in blossom from the beginning of May till the end of July.

The juice of this plant is acrimonious: it is said to cure ring-worms, and, when diluted with milk, to consume white opaque spots in the eyes. It is also thought efficacious in the cure of warts and cutaneous disorders. The root is esteemed by the natives of Cochin-China for a variety of medicinal purposes.

This species preserves its green leaves all the year, and they are remarkably handsome; being large, elegantly shaped, and of a transparency which shows the delicacy of their texture, as the yellow light shines through them. The double-flowered variety is chiefly cultivated in gardens: it is increased by parting the roots in autumn.

The usual mode of sowing these plants is to scatter the seeds about in rock-work, where they will come up without further trouble. If sown in pots, the best time for the purpose is in September: one seed in each pot. They

* This is the proper swallow wort; and called, in *French,* l'eclaire, la grande eclaire, le felongéne, l'herbe de l'hirondelle [swallow's herb:] in *Italian,* favagella, cerigogna.

should stand in the open air, and they require watering only in very dry weather: the last-mentioned species loves the shade.

The Small Celandine, or Pilewort, is not usually admitted into gardens; but, on the contrary, on account of the injury it does to every thing growing near it, is carefully rooted out wherever it appears. It is a species of ranunculus, called the ranunculus ficaria, from the shape of the root, which resembles that of the fig; and belongs to the natural family of the Ranunculaceæ.

In early spring, there is scarcely a grove, thicket, meadow, hedge, orchard, or plantation of any kind, that is not covered with the glossy golden flowers of the Small Celandine. When they have been exposed for some days to the heat of the sun, they turn white, and fall off: they are succeeded by small bulbs, like grains of wheat, which shoot from the bosom of the leaves; and as the stalks lie upon the ground, these little bulbs get into the earth, and become the roots of new plants. The stalks being sometimes washed bare by the rains, have induced the ignorant and superstitious to believe that it rained wheat. The young leaves are eaten by the common people of Sweden, boiled as greens.

At night, and in wet weather, the flowers close, which helps to preserve them from the cold that otherwise might be hurtful to them, from their flowering so early in the spring. They first appear in February, and continue through March, and a great part of April. It seems, the early flowering of this plant has helped to recommend it to the notice of Mr. Wordsworth, by whom it has been highly and repeatedly celebrated:

"Pansies, lilies, kingcups, daisies,
Let them live upon their praises;

Long as there's a sun that sets,
Primroses will have their glory;
Long as there are violets,
They will have a place in story:
There's a flower that shall be mine,
'Tis the little Celandine.

Eyes of some men travel far
For the finding of a star;
Up and down the heavens they go,
Men that keep a mighty rout!
I'm as great as they, I trow,
Since the day I found thee out,
Little flower!—I 'll make a stir
Like a great astronomer.

Modest, yet withal an elf,
Bold, and lavish of thyself,
Since we needs must first have met
I have seen thee, high and low,
Thirty years or more, and yet
'Twas a face I did not know;
Thou hast now, go where I may,
Fifty greetings in a day.

Ere a leaf is on a bush,
In the time before the thrush
Has a thought about its nest,
Thou wilt come with half a call,
Spreading out thy glossy breast
Like a careless prodigal;
Telling tales about the sun,
When we've little warmth, or none.

Poets, vain men in their mood!
Travel with the multitude;
Never heed them; I aver
That they all are wanton wooers;
But the thrifty cottager,
Who stirs little out of doors,
Joys to spy thee near her home,
Spring is coming, thou art come!

Comfort have thou of thy merit,
Kindly, unassuming spirit!
Careless of thy neighbourhood,
Thou dost show thy pleasant face
On the moor, and in the wood,
In the lane—there 's not a place,
Howsoever mean it be,
But 'tis good enough for thee.

Ill befal the yellow flowers,
Children of the flaring hours!
Buttercups, that will be seen,
Whether we will see or no;
Others, too, of lofty mien;
They have done as worldlings do,
Taken praise that should be thine,
Little, humble Celandine!

Prophet of delight and mirth,
Scorned and slighted upon earth;
Herald of a mighty band,
Of a joyous train ensuing,
Singing at my heart's command,
In the lanes my thoughts pursuing,
I will sing, as doth behove,
Hymns in praise of what I love."

But to quote all this poet's praises of the Celandine is more than can be allowed to us. The reader is too well acquainted with his writings to be ignorant of his love for this little flower, or to refuse him the sympathy he requires:

"Let, with bold advent'rous skill,
Others thrid the polar sea;
Build a pyramid who will;
Praise it is enough for me,
If there be but three or four
Who will love my little flower."

Mrs. Charlotte Smith more than once alludes to the

early flowering of the Pilewort; particularly in the lines addressed to the early butterfly:

"Trusting the first warm day of spring,
When transient sunshine warms the sky,
Light on his yellow spotted wing
Comes forth the early butterfly.

With wavering flight he settles now
Where Pilewort spreads its blossoms fair,
Or on the grass where daisies blow,
Pausing, he rests his pinions there.

But, insect, in a luckless hour
Thou from thy winter home hast come,
For yet is seen no luscious flower,
With odour rich and honied bloom.

And these that to the early day
Yet timidly their bells unfold,
Close with the sun's retreating ray,
And shut their humid eyes of gold."

CENTAURY.

CENTAUREA.

CINAROCEPHALEÆ. SYNGENESIA POLYGAMIA FRUSTANEA.

This plant has been also named Chironium; both names being derived from the centaur Chiron; some say, because first discovered by him—others, from his having been cured by it of a wound in his foot, made by the fall of an arrow when he was entertaining Hercules.—*French,* la centaurée; bluet; barbeau; aubifoin.—*Italian,* centaurea.

THIS is a very extensive genus, greatly varying in beauty: some being mere ordinary weeds, others handsome and showy flowers. Many of them are cultivated in our gardens: the most common, perhaps, is the Sultan-flower, or Sweet-sultan, a native of Persia, and commonly seen growing wild among the corn in the Levant. The colour is purple, flesh-coloured, or white. The scent is very powerful, and to some persons disagreeable.

There is a variety, called, from the colour of its flowers, Yellow Sweet-sultan*, of which the scent is unquestionably pleasant. The best time to sow Sweet-sultan is in the spring: they will begin to flower in July. One seed will suffice for a six-inch pot: water must be given sparingly, or the roots will be liable to rot. The yellow variety is raised in a hot-bed, and, when grown, requires more tender treatment than the rest of the family. They are annual plants.

The perennial kinds may be either increased by seed, as directed, or by parting the roots in autumn: always observing to place such as are newly planted in the shade until they have taken fresh root. These will require shelter in the winter. Centaury has a tendency to strike very deep root, which makes many of them altogether unfit for pots. Unfortunately, the Great Centaury is of this number: I say unfortunately, because this species, which grows naturally on the mountains of Italy, has been rendered classical by Virgil's mention of it in his Georgics, where it is recommended, among other flowers, as a medicine for bees when sick. I think Dryden also mentions it somewhere.

* The centaurea amberboi of the botanists. In French, *le barbeau jaune; fleur du grand seigneur; l'amberboi.*—Italian, *ciano giallo Turchesco odoroso.*

CEREUS.

CACTUS.

OPUNTIACEÆ. ICOSANDRIA MONOGYNIA.

The origin of the name uncertain.—*French,* le cactier.

The Great-flowered Creeping Cereus, called in French *le serpent,* is a plant of extraordinary magnificence and beauty. Its blossoms open in the evening: they are large and sweet-scented, but of very short duration. They begin to open between seven and eight o'clock; are fully blown by eleven, and by three or four in the morning they fade, and hang down quite decayed. During their short-lived beauty, few flowers can compare with them. The calyx of the flower, when open, is nearly a foot in diameter; the inside of which, being of a splendid yellow, appears like the rays of a bright star: the outside is of a dark brown. The petals of the flower are of a pure and dazzling white; and a vast number of recurved stamens, surrounding the style in the centre, add to its beauty. The fine scent of this extraordinary flower perfumes the air to a considerable distance. It flowers in July; and upon large plants eight or ten flowers will open on the same night, and be succeeded by others for several nights together, making a most magnificent appearance by candle-light. This plant does not bear fruit in this country, and must be nursed in a stove, to enable it to produce flowers. It is, in fact, an intruder here; but it is to be hoped its beauty will obtain pardon for its intrusion: the more readily, as it introduces a very lovely relation, who has right of admission.

The Pink-flowered Creeping Cereus produces a greater number of flowers than the former. They open in May, or, in warm seasons, yet earlier. They are of a fine pink colour, and keep open three or four days. This plant has very slender branches, which should be trained to a little trellis frame of sticks. The flowers are so beautiful and so numerous, that it deserves some care to cherish it. It may be preserved through the winter in a warm inhabited room, and towards the end of May may be set abroad. Very little water must be given in summer, and scarcely any in the winter. About the middle of September it should again be removed into the house. If there be much rain or sharp winds in the summer season, this plant must be sheltered; and it must always be in a warm situation. It will flower better if it can conveniently be placed within the room even in summer, if near to an open window. It should not have a very large pot, or a rich soil. This plant is a native of Peru: the former species, of Jamaica.

The Six-angled Upright Cereus, or Torch-thistle—in French, *le cactier de Surinam*—was the first which became common in English hot-houses. This plant, if not cut down, will grow forty feet high; but wherever the stems are cut, they put out others from the angles immediately below the wounded part. The flowers are white, and as large as those of the hollyhock. It does not often flower; when it does, it is generally in July. It is a native of Surinam, and may be preserved in the same manner as directed for the Pink-flowered species. The cochineal insect feeds chiefly upon plants of this genus, and the Indians frequently propagate them for the sake of those insects; particularly that which is called the Cochineal Indian-Fig.

CERINTHE.

HYDROPHYLLEÆ. PENTANDRIA MONOGYGNIA.

Cerinthe is derived from the Greek, and signifies honeycomb, which, as well as the name of honeywort, has been given to this plant on account of the quantity of honey-juice it contains.—*French,* le mélinet. *Italian,* cerinte or cerinta.

THE Great Honeywort has a purple flower, with a yellow tube: the Small, a yellow flower. They will continue in blossom the greater part of the summer. As it injures the seeds to remain long out of the ground, they should be sown in autumn, soon after they are ripe: sow the seeds singly, in four or in five inch pots; house them during frost, and keep them moderately moist. They are both annual plants; pretty while they last, and of an agreeable scent. The honey-juice contained in the tube of the flowers is a great attraction to bees; and it is for this reason recommended as proper to plant near apiaries.

Virgil recommends the keepers of bees to sprinkle the fragrant juices of Balm and Honeywort, to entice them home. Cerinthe is one of the most common herbs in the fields of Italy; which induces Virgil to term it *ignobile gramen:*

"Huc tu jussos asperge sapores,
Trita melisphylla, et cerinthæ ignobile gramen."

Dryden translates *melisphylla* and *cerinthe,* melfoil and honeysuckle:

"Then melfoil beat, and honeysuckles pound,
With these alluring savours strew the ground."

But we have no plant named Melfoil. Milfoil is so called from its great number of leaves. Rucellai, in his Italian poem, translates the passage thus:

"E però sparga quivi il buon sapore
De la trita melissa, e l'erba vile
De la cerinta."

LE API DEL RUCELLAI.

And therefore sprinkle here the genial flavour
Of the bruised balm and lowly honeywort.

CHELONE.

BIGNONIEÆ. DIDYNAMIA ANGIOSPERMIÆ.

This name is derived from the Greek, and signifies a tortoise.—*French,* galane; tortue, [tortoise].

THE White Chelone has been called by Joscelin, in his New England Rarities, the Humming-bird Tree. When planted in the open ground, it spreads its roots to a considerable distance; but it rather improves than injures them to confine the roots by putting the plant in a pot; as the stalks which the root sends up will otherwise be too far distant, and have a straggling appearance.

The Red Chelone is very similar to the first species, but has broader leaves, and the flowers being of a brilliant purple, it is altogether more showy than those with white flowers.

The Hairy Chelone is also very similiar to the first, but that the leaves of this are hairy, and the flowers are of a clearer white.

There are one or two others, but these are the handsomest; and as their treatment should be the same, it is useless to make a mere catalogue of names.

They are all natives of North America, and will endure the cold without injury, but must be watered daily in hot weather; and, when very dry, both morning and evening. These plants are the more valuable, as they are in full beauty in the autumn, when most flowers are beginning to decay.

CHIONANTHUS.

JASMINEÆ. DIANDRIA MONOGYNIA.

The name of this shrub is derived from the Greek, and signifies snow-flower. It is usually called the Virginian snow-drop tree.—*French,* l'arbre de neige.—*Italian,* albero di neve.

THIS shrub is common in South Carolina, where it grows by the side of rivulets. The flowers come out in May, hanging in long bunches, and are of a pure white; whence it is called by the inhabitants Snowdrop Tree: and, from the flowers being cut into narrow segments, they give it also the name of Fringe Tree.

The Snowdrop Tree requires much care in raising: the best time to procure one is when it is about four years old; it will then endure the cold of winter. In the summer it likes the morning sun; and is always fond of water. In dry summer weather it may be refreshed with a little water, both morning and evening.

CHIRONIA.

GENTIANEÆ. PENTANDRIA MONOGYNIA.

This genus, like the centaury, is named after the centaur Chiron.

THERE are several species of Chironia, which, being chiefly natives of the Cape, may be treated in the same manner. They are little shrubby plants, varying in colour according to the species: blue, purple, yellow, or red. The most common are the berry-bearing kinds, of which there are two; one, which is, on this account, named the Berry-bearing Chironia; the other, Frutescens, or Fruit-bearing. The first of these is both in flower and in fruit during nearly the whole of the summer.

These plants must be housed in the winter, but so placed as to receive as much sun as possible; and fresh air in mild weather. They must be observed daily, that they may not be left with the earth dry, but must have only water sufficient to prevent this, particularly in the winter; and must be preserved from damps.

CHRYSANTHEMUM.

CORYMBIFERÆ. SYNGENESIA POLYGAMIA SUPERFLUA.

This name is derived from the Greek, and signifies gold-flower.

THIS article will be found to contain some of the Marygolds, of which the different kinds are so dispersed, and so intimately connected with many different genera, that it would rather increase than lessen the confusion to place them all under one head. The Index will refer to such articles as relate to them.

One of the handsomest of the Chrysanthemums is the Indian; the flowers of which are three inches or more in diameter: it varies in colour; there are white, purple, red, orange, yellow, &c. This kind requires shelter in the winter; as also does the Canary Ox-eye, a native of the Canary Islands, very much resembling the common chamomile flower. In winter, these two kinds should be very gently watered, about three or four times in a week: in the summer, they will require it more plentifully, and every evening when the weather is dry.

The Siberian Chrysanthemum is very hardy, and will live in the open air all the year: it does not often perfect seeds in England, but may easily be increased by slips, which may be planted two or three in a pot, in September or October, and transplanted into separate pots in March; it will be necessary to shelter these young shoots in frosty weather, and to keep the earth moist.

The Garden Chrysanthemum, sometimes called the Cretan, or Cretan Corn Marygold, is yellow; it flowers in June. This is an annual plant, and generally raised in a hot-bed. It is not, however, very tender; and cuttings planted in autumn, and kept in the house in the winter, will, if in a tolerably warm situation, take root, and flower well in the summer.

The common Ox-eye, likewise called Ox-eye Daisy, Maudlin-wort, and Moon-flower, is a perennial plant, very common in dry pastures, corn-fields, &c. It is called in French, *la marguerite grande* [great daisy]; *la grande paquette; l'œil de bœuf* [ox eye]; *l'œil de bouc* [goat's eye]: and in Italian, *leucantemo* [white flower]; *la margheritina maggiore* [great daisy]; *l'occhio di bue* [ox eye]. The flower is white, with a yellow eye. It has been much recommended for its medicinal virtues, but does not appear to have established its reputation in this respect: the young leaves are eaten in salads; and it is said are, in Padua, much esteemed for this purpose. It continues in blossom from May till July; will live in the open air; and should, as well as all the other kinds, be kept moist.

There are several other species, which generally require the same treatment; that is, moderate watering, and winter shelter.

The common Corn Marygold, which belongs to this genus, known in France by the name of *la marguerite jaune* [yellow daisy]; *souci des champs* [field marygold]; *souci des blés* [corn marygold]; and in Italy, by those of *crisantemo* [gold flower]; and *margherita gialla* [yellow daisy]; is seldom grown in gardens: it is very common in corn-fields; and, as Linnæus observes, though their brilliant colours may please the eye of the passing traveller, they are no very agreeable sight to the farmer, to whom they are but troublesome weeds. He informs us, that there is a law in Denmark to oblige the farmers to extirpate them. These flowers are also called Gowans, Gules, Gools, Gowls, Guills, Goulans, Goldins, Yellow-bottles, and Golden Corn-flowers. The Germans use them as a yellow dye. The Chrysanthemum, the Indian particularly, is in high estimation with the Chinese, and is celebrated by all their poets *.

CINERARIA.

CORYMBIFERÆ. SYNGENESIA POLYGAMIA SUPERFLUA.

Ash-coloured; most of the species being of a grayish colour.—*French*, cendriette; cinerre.—*Italian*, cineraria.

The handsomest kinds are the Blue-flowered Cineraria, or Cape-Aster, and the Woolly Cineraria. The flowers of the first are of a bright sky-blue, and the plant is never without them the whole year round. Of the second, the inner part of the flowers is white, the outside a most vivid purple: it flowers early in the spring, and, if in a healthy state, will also flower all the year; but this plant is often infested with a kind of insect which destroys its vigour; therefore, to ensure a succession of healthy, handsome plants, it should be annually increased by cuttings, which, if planted in September, and placed in a tolerably warm situation, will strike root very readily.

These plants must be housed in the winter. Many persons keep the last kind in a stove, but, like many of ourselves, they are more healthy when treated less tenderly. The earth must be kept moderately moist.

* See Titsinghi's Illustrations of Japan.

CISTUS.

CISTEÆ. POLYANDRIA MONOGYNIA.

Called also gum cistus, and rock rose.—*French,* le ciste.—*Italian,* cisto, cistio.

The Cistus is a very extensive genus, and all the species are valuable ornaments to a garden. Their flowers, although of short duration, are succeeded almost every day by fresh ones, for more than two months, and are generally about the size of a rose. They are of different colours, and the plants retain their leaves all the year.

Some few require a stove; it will be sufficient to specify the most beautiful kinds which may be preserved without one.

The Poplar-leaved Cistus, a native of Portugal: flowers white, tinged with purple at the edges; bloom in June and July.

The Bay-leaved Cistus, a native of Spain: flowers white; blow in June and July.

The Spanish Gum Cistus: white flowers, with spots of purple at the base; in blossom from June to August.—The whole plant exudes a sweet glutinous substance in warm weather, which has a strong balsamic scent, and perfumes the air to a great distance.

The Montpelier Gum Cistus, a native of Narbonne and Valencia: white flowers, open from June to August.—This species exudes a gum, like the last. There is a variety of it, with lemon-coloured flowers.

The Hoary Rock Rose, or Rose Cistus, *le ciste ordinaire* of the French: a native of Spain and Narbonne; purple flowers.

The Cretan Cistus: a native of the Levant; flowers red purple, blowing in June and July. This is frequently called the Ladaniferous Cistus, being that from which the drug called ladanum is obtained: a kind of resin, which, on account of its fragrant smell, is frequently used in fumigations.

The White-leaved Cistus, a native of Spain and Narbonne: flowers purple. June and July.

The Sea Purslane-leaved Cistus, a native of Portugal; with large bright yellow flowers, which appear in June and July.

These Cistuses are shrubs, from one foot to five or six feet high. They must be housed at the approach of winter, and gradually replaced in the open air early in the spring. The earth should be kept moderately moist.

The Dwarf Cistus, or Little Sunflower, is an indigenous plant: it is called in France, *la fleur du soleil* [sun flower]; *l'hysope des Carigues; l'herbe d or* [golden herb]: and in Italy, *eliantemo* [sun flower]; *fior del sole* [sun flower]. The flowers are usually a deep yellow, or pale lemon colour; but they are sometimes seen white, and rose-coloured.—All these varieties, placed together, have an agreeable effect. This species will live in the open air, all the year round.

CLEMATIS.

RANUNCULACEÆ. POLYANDRIA POLYGYNIA.

Called frequently, virgin's bower, or traveller's joy.—*French,* l'herbe au gueux [beggar's herb]; la viorne; viorne des pauvres [poor man's rest]; la consolation des voyageurs [traveller's consolation]; in the villages, vouabla, a corruption of the Latin name vitalba [white vine].—*Italian,* vitalba; clematite.

THESE are, for the most part, climbing plants, needing support, and should be placed where they may run up a wall or balcony. They will not flower so strongly in pots as in the open ground; but must not, on this account, be rejected. The Evergreen Clematis would require to be planted in a tub of some magnitude: it grows to the height of eight or ten feet, and becomes very thick and bushy. The flowers are of a greenish colour, and appear in December or January. It retains its leaves all the year.—Gerarde gives it the name of Traveller's Joy of Candia; Johnson, Spanish Traveller's Joy; and Parkinson, Spanish Wild Climber.

Purple Clematis grows naturally in the woods of Spain and Italy: there are several varieties, the Single Red-flowered, Blue-flowered, and Purple-flowered, and the Double Purple; which flower, in June, July, and August: and another with white flowers, which appear in May.—Gerarde gives this species the name of Climbing Ladies' Bower, "from its aptness," he says, "to make bowers or arbours in gardens."

The Curled Clematis is a native of Carolina, Florida, and Japan; the stalks grow near four feet high, and fasten themselves by their claspers or tendrils, to the neighbouring plants. The flowers are purple, and blow in July.

The Oriental Clematis is a native of the Levant; it has flowers of a greenish yellow colour, which are in blossom from July till October.

The Upright Virgin's Bower, or Clematis Flammula, (in French, *la flammule; clematite odorante:* Italian, *flammula:*) grows naturally in many parts of Europe. The flowers are white, and continue in blossom from June till September. This is an acrid, corrosive plant, and inflames the skin, whence it has been named Flammula.

The Hungarian Clematis has blue flowers, which are in blossom from June to August. This and the last mentioned species have annual stems.

All the kinds here enumerated, which are the handsomest, will live in the open air all the year. They should, in general, be watered about three times in a week, but in very hot and dry weather every evening.

There are some few species of the Clematis which require artificial heat, but they are by far the least handsome. The two last mentioned kinds may be increased by parting the roots, which should be done either in October or February. The roots may be cut through their crowns with a sharp knife, taking care to preserve some good buds to every off-set.

The Clematis is as great a rambler as the Honeysuckle itself:—

"o'errun
By vines, and boundless clematis, (between
Whose wilderness of leaves, white roses peep'd)
And honeysuckle, which, with trailing boughs,
Dropp'd o'er a sward, grateful as ever sprung
By sprinkling fountains."

BARRY CORNWALL.

Mr. Keats makes mention of the Clematis in a passage, of which, as it relates entirely to flowers, it may, perhaps, be allowable to quote the whole. He describes a youth sleeping in a bower walled with myrtle:

"Above his head
Four lily-stalks did their white honours wed,
To make a coronal, and round him grew
All tendrils green, of every bloom and hue,
Together intertwined, and trammel'd fresh:
The vine of glossy sprout; the ivy-mesh,
Shading its Ethiop berries; and woodbine
Of velvet leaves, and bugle blooms divine;
Convolvulus in streaked vases flush;
The creeper, mellowing for an autumn blush;
And virgin's bower, trailing airily,
With others of the sisterhood."

ENDYMION, p. 72.

CLETHRA.

ERICINEÆ. DECANDRIA MONOGYNIA.

The Clethra Arborea, or Tree Clethra, will require shelter from the winter cold, in our climate: it should be housed about the middle or end of September, according as the weather is more or less mild; and, during this season, should be watered about twice a week; in the summer, when the weather is dry, it should be watered once in a day, or in two days, in proportion to the heat of the sun, or the plant's exposure to it. The earth should not be suffered to become parched. It is a native of Madeira.

COLCHICUM.

COLCHICACEÆ. HEXANDRIA MONOGYNIA.

So called from Colchis, a city of Arminia, where this plant is supposed to have been very common. The English name of meadow saffron is from its common place of growth, and its resemblance to the crocus, or saffron flower.

The Autumnal Colchicum, or Common Meadow Saffron, is named in French, *tue chien*, *mort au chien*, both signifying dog poison; in the villages, *bovet;* in Italian, *colchico;* and has many varieties: the Yellow-flowered or Crocus Colchicum, the Purple, Red, White, Rosy, Rosy-variegated, Purple-variegated, and Double. The flowers appear in autumn, the leaves not till the following March; for which reason the country people call them Naked Ladies, an appellation bestowed upon many flowers which blow before they are in leaf.

There are several other species, requiring the same treatment as this. The roots are bulbous, and a new one is formed every year, as the old one decays. The leaves begin to wither in May, soon after which the roots should be taken out of the earth, put in a shady place to dry, wiped clean from earth, decayed fibres, &c. and put into a dry place, safe from insects, &c. until the beginning of August, when they should be planted again, about three or four inches deep, in a sandy soil.

The pot should be about six inches wide and nine deep. Water should be given in small quantities, and if the pot be placed in the shade, exposed to the dews and light summer showers, it need not be watered at all, until after the plant has begun to shoot above the earth.

It injures the root of the Colchicum to pluck the flower when newly blown, as it deprives the new root which is forming of a part of its nourishment. It will likewise be improper to delay planting the roots after the beginning of August, as they will otherwise vegetate, and produce their flowers without planting, which will greatly weaken them.

COLUMBINE.

AQUILEGIA.

RANUNCULACEÆ. POLYANDRIA PENTAGYNIA.

Cock's-foot or culverwort.—The botanical name for this plant, Aquilegia or Aquilina, is derived from aquila, an eagle, from a notion that the nectaries resemble an eagle's claws. Our English name, columbine, is derived from the resemblance which, in a wild state, these parts bear, both in form and colour, to the head and neck of a dove, for which the Latin name is columba.—*French*, aiglantine, la colombine, la galantine; gands de notre dame [our lady's gloves].—*Italian*, achellea, colombina, perfetto amore [true love], celidona maggiore [great celandine]; at Venice, galeti.

The Common Columbine is generally, in its wild state, of a blue colour, whence it is named the Blue Starry, but in the neighbourhood of Berne, and in Norfolk, it has been found both with red and white flowers. It is common in woods, hedges, and bushes, in most parts of Europe. They are greatly changed by culture; become double in various ways; and are of almost all colours; blue, white, red, purple; flesh, ash, and chestnut coloured; blue and white, and red and white. It is a perennial plant, and, with us, flowers in June.

Every part of this plant has been considered as a useful medicine, but Linnæus affirms that, from his own knowledge, children have lost their lives by an over dose of it. That might, however, be the case with some of our best medicines.

The Alpine Columbine has blue flowers tipped with a yellowish green, blowing in May and June. (Biennial).

The Canadian Columbine flowers in April: the flowers are yellow on the *in*, red on the *out*side. (Perennial).

The Columbines may be increased by parting the roots; but, as they are apt to degenerate, are most commonly raised from seed: these will not grow to flower till the second year; and, as you cannot be sure of the kinds they will produce, it is better to procure the plants from a nursery. They should have a little water, two or three times a week, in dry weather; and may remain in the open air.

Gawin Douglas speaks of the Columbine as black, from the deep purple which some of them take:

> "Floure-damas, and columbe blak and blew."

This has been differently expressed in Mr. Fawkes's modernized version; and not happily, for the Columbine *drops* its head:

> "And columbine advanced his purple head."

W. Browne speaks of it in all its colours:

> "So did the maidens, with their various flowers
> Decke up their windowes, and make neat their bowers;

Using such cunning, as they did dispose
The ruddy piny with the lighter rose,
The monk's-hood with the bugloss, and intwine
The white, the blewe, the flesh-like columbine
With pinks, sweet-williams; that, far off, the eye
Could not the manner of their mixtures spye."

He tells us that the King-cup is an emblem of jealousy; that—

"The columbine in tawny often taken,
Is then ascribed to such as are forsaken;
Flora's choice buttons, of a russet dye,
Is hope even in the depth of misery;
The pansie, thistle all with prickles set,
The cowslip, honeysuckle, violet,
And many hundreds more that grace the meades."

A preparation from the Columbine has been administered to children, in the same manner as the Syrup of Poppies; and Linnæus says he has seen them die in consequence.

COLUTEA FRUTESCENS.

LEGUMINOSÆ. DIADELPHIA DECANDRIA.

Usually called Scarlet bladder-senna.

THIS shrub is a native of the Cape: the flowers are of a fine scarlet, and, intermingled with its silvery leaves, are very handsome. If the plant is treated hardily, it seldom lives more than two years; but it is much handsomer and fuller of fower while it does last than such as are treated in a more tender manner. The best way to manage it, is to let it remain abroad altogether, till the middle, or, if tolerably mild, till the end of October. It should then be housed at night, but placed near to an open window, and put abroad, in as warm a situation as can be chosen for it, in the day-time, whenever the weather is not frosty. On frosty days it should remain in its night's lodging. When the frosts are securely over, it may be again left altogether in its out-door station. The flowers appear in June. The earth should be kept moderately moist.

CONVOLVULUS.

CONVOLVULACEÆ. PENTANDRIA MONOGYNIA.

Commonly known, when wild, by the name of bind-weed, from some of the species twining their stem round other bodies, which is also the signification of the Latin name.—*French*, le liseron.—*Italian*, il villuchio.

THIS is a most extensive genus: Martyn's edition of Millar's Dictionary mentions 110 different species, besides a great many flowers of different genera, which are intimately connected with it.

The Common Field Bind-weed is one of the greatest pests to gardeners and farmers. It is yet worse than the Hedge Bind-weed; for that, for the sake of climbing, confines its ravages to the borders of the fields or gardens, while this wanders over the whole ground, and is with great difficulty rooted out. And yet it must be acknowledged that this little red and white flower is extremely beautiful; and, were it but a little more modest, would, doubtless, be a general favourite. As it is, it must suffer the consequence of its impertinence, not only in being avoided, but positively turned out. From the frequent occurrence of this beautiful intruder, it has acquired a multitude of names, as bell-bind, bell-wind, rope-weed, with-wind. In French, *la lizeret, le liseron des champs;* in Provence, *courregeolo;* in Languedoc, *campanette;* in Lorraine, *oeillet* [pink]. In Italian, *vilucchio, viticchio; correggiola; campanella; convolvolo:* in the Venetian territories, *broeca:* in the Brescian, *tirangolo.*

There are comparatively few of these plants cultivated in our gardens. The following are some of the most esteemed.

The Two-coloured; white and purple, flowering in June, July, and August. The Hairy Convolvulus, with purple flowers, blowing at the same time. These are natives of the East Indies.

The Five-petaled; blue, with a yellow centre: native of Majorca. Flowers from June to August.

The Indigo Convolvulus, which is named from the colour of its flowers: it is a native of America, and considered one of the handsomest of the genus. The Italians call it *campana azurea* [azure-bell], and *fior di notte* [night-flower], because its beauty appears most at night. It blows in July and August.

Of the Major Convolvulus there are three or four varieties; purple, white, red, and pale blue. It is a native of America. It requires support, and will grow ten or twelve feet high; continuing in flower from the beginning of June till the approach of frost.

The Minor Convolvulus is a native of Spain and Portugal; the flowers are sometimes pure white, but more commonly variegated with blue and yellow, or blue and white: the most beautiful kind is a bright blue, fading, by delicate gradations, to a pure white in the centre. It resembles the blue atmosphere, relieved by fleecy clouds, on a fine day in summer:

"when on high,
Through clouds of fleecy white, laughs the cerulean sky."

KEATS.

Nor is the form of this flower less beautiful than the colour, either when spread out in full beauty to the midday sun, or when, at the approach of night, it closes its blue eye to sleep.

This flower is too well known to need description; but its exquisite loveliness impels one to linger over it with admiration.

All the kinds here specified are annual plants. The Five-petaled, the Major, and the Minor, may be raised at home with little trouble. The seeds may be sown about an inch asunder. As some may fail, they may at first be scattered more closely; and, as they come up, thinned where they crowd each other. If sown in the autumn, they will flower in May: those sown in spring will be a month later. They may be sown in September and March; and, for a longer succession, in April and May likewise. The other kinds must be raised in a hot-bed, and will not bear the open air in the winter.

The Dwarf Convolvulus is a native of France, Spain, and Sicily. It has deep rose-coloured flowers, is a perennial plant, and will live in the open air. It may be increased by parting the roots, either in spring or autumn.

The Canary Convolvulus, with pale blue or white flowers, blowing in June and July, is a native of the Canary Islands.

The Silvery Convolvulus, with pale rose-coloured flowers, opening in June, July, and August, is found in Spain, Sicily, the Levant, &c.

The Arabian—but there will be no end of enumeration at this rate. The Canary and Silvery kinds must be housed in the winter. With respect to the variegated kinds, if a plain flower appear, care should be taken to pluck it immediately, in order to prevent the succeeding blossoms also from degenerating from their natural beauty. The earth should be kept moderately moist, and the water given in small quantities at one time. The plants, being mostly tall and slender, should be sheltered from heavy beating rains and violent winds; but light spring or summer showers will refresh them.

This genus furnishes to the materia medica two of its most powerful drugs: scammony, from a species growing naturally at the Levant; and jalap, from another kind, which is a native of Xalapa, between Vera Cruz and Mexico. They are obtained from the roots of the plants. Most of these flowers close at night; and many remain close all day when the weather is wet or cloudy, but open to the sunshine:

> "Qual' i fioretti dal notturno gielo
> Chinati e chiusi, poi che 'l sol gl' imbianca,
> Si drizzan tutti aperti in loro stelo."
>
> DANTE, INFERNO, Canto II.

> "Like flow'rs, which shrinking from the chilly night,
> Droop and shut up; but with fair morning's touch
> Rise on their stems, all open and upright."

COREOPSIS.

CORYMBIFERÆ. SYNGENESIA POLYGAMIA FRUSTANEÆ.

The generic name is from the Greek, bug-like, the seed being like a bug or tick: hence it is called by gardeners the Tick-seeded Sunflower.

THE Whorl-leaved Coreopsis has a yellow flower with a purple centre: it is a showy plant, grows very tall, and continues long in flower. It begins to blossom in July. It is a native of North America, where the flowers, although yellow, are used to dye cloth red.

The Three-leaved has the same coloured flowers, and is from the same country.

The Alternate-leaved, Thick-leaved, and Golden, are all from North America. The first flowers in October and November; the other two from August to October. These are all perennial plants, as are most of the genus.

They may be increased by parting the roots, which should be done in autumn, when the stalks begin to decay. The two first prefer a light loamy earth, and exposure to the sun; the others will thrive in almost any soil or situation. There are other species of this genus, some of which are raised in a hot-bed; but their treatment, when grown, is generally the same. The kinds here named will bear the open air. The earth should be kept just moist, and the plants be supported by sticks as they advance in height, or the strong winds of autumn may be apt to break them.

CORN-FLAG.

GLADIOLUS.

IRIDEÆ. TRIANDRIA MONOGYNIA.

The botanical name of this plant is the diminutive of *gladius*, a sword, and is given it from the form of its leaves. It is also called Sword-flag, Corn-sedge, and Corn-gladin.—*French*, le glayeul; flambe. —*Italian*, ghiaggiuolo; gladiolo.—In Sicily, spatulidda.

The Corn-flag is related to the lily, and has a bulbous root. It is a handsome genus. Of the Common Corn-flag there are many varieties, differing in colour. These may be increased by offsets from the roots. About the end of July, when the stalks decay, the roots may be taken up, the offsets separated from them, and the whole dried, cleaned, and carefully preserved in a dry and secure place till the end of September or the beginning of October, when they may be re-planted. They will bear the open air.

The other species are chiefly natives of the Cape, and require this difference in their treatment, that they must be kept within doors from October till May, allowing them fresh air in mild weather.

The Corn-flags must be sparingly watered; in the winter, not more than once a week. The roots should be planted separately, in pots about five inches wide, and should be covered two inches deep.

CORONILLA.

LEGUMINOSÆ. DIADELPHIA DECANDRIA.

The name of this plant is derived from *corona,* a crown, of which it is the diminutive; the flowers crowning the branches in a cluster.

THE Coronilla Emerus, or Scorpion Senna, is a native of most parts of the Continent of Europe. The flowers are yellow, and blow in April. A dye is obtained from this plant nearly equal to that of indigo.

This shrub is fond of water.

The Small Shrubby Coronilla has small deep yellow flowers, blowing in May, June, and July. It has a very powerful scent, and is a native both of Spain and Italy.

The Great Shrubby Coronilla is very similar to the last; but this is in flower almost all the year; and the scent of it is more powerful in the day-time than in the night. It is a native of the South of France.

The Cretan Coronilla is a very low shrub, but very handsome when in full blossom, as it produces an abundance of yellow flowers. They blow in May, and are very sweet scented.

The three last kinds are not so fond of water as the first, but incline to a dry soil. In dry summer weather they may be slightly watered about three times a week. In the winter they should be sheltered from the frost, and then once in a week will suffice to water them. This treatment will suit most of the species.

COTYLEDON.

CRASSULACEÆ. DECANDRIA PENTAGYNIA.

Called also Navelwort, which is the signification of the botanic name in the Greek.

The Round-leaved, Oval-leaved, and Oblong-leaved, are properly only varieties of the same species. They are natives of the Cape, and are in blossom from July to September. They must be sheltered in the winter. They are extremely succulent; and care must be taken to preserve a due medium in watering them. If they have too much wet, it will rot them: too little will not nourish them. Observe the leaves, and do not let them shrink for want of moisture. Give them just sufficient to keep their vessels distended. It must be shed on the roots only.

There are many species of Navelwort. Those which do not require a stove may be treated in the same manner as those already mentioned. They are all very succulent, and should have a poor, dry soil. They may be sown either in spring or autumn.

The flower called Venus's Navelwort has no affinity with these, but is the cynoglossum linifolium. It is an annual plant. The seeds may be sown pretty thick, either in spring or autumn; and, if they all grow, they should be thinned where too close. Those sown in autumn will flower in May and June. The spring-sown seeds will come to flower a month later. The earth should be moderately moist.

COWSLIP.

PRIMULA VERIS.

PRIMULACEÆ. PENTANDRIA MONOGYNIA.

The Cowslip, *i. e.* cow's lip, is of the same genus as the primrose. The Yorkshire people call the Cowslip Cow-stripling. It is also called Herb-Peter, and Paigles.—*French,* la primevère, primerole; herbe de la paralysie [palsy herb]; fleur de coucou; bavillon.—*Italian,* primavera.—In the Venetian territory, primola.—Some of these are also used for the primrose.

THE Common Cowslip, or Paigle, is common in Europe, both in moist sand and upland pastures, and on the borders of fields. In a clayey or loamy soil it thrives best, and prefers an open situation. It flowers in April and May. Though respected both for its beauty and utility, the Cowslip, in pastures where it is very common, becomes an injurious weed. The leaves are eaten in salads, and recommended for feeding silk-worms before the mulberry-leaves make their appearance. The flowers are very fragrant; and a pleasant and wholesome wine is made from them, approaching in flavour to the muscadel wines of the South of France. It is said to be an inducer of sleep.

——————"For want of rest,
Lettuce and cowslip-wine: *probatum est.*"

POPE.

These flowers have a rough and somewhat bitter taste, which, with their agreeable odour and yellow colour, they impart both to water and spirit. A pleasant syrup is made from them; and a strong infusion, drank as tea, is considered antispasmodic. The colour, as is well known, is usually a bright yellow, dashed with deep orange, sometimes approaching to crimson.—Thus Iachimo describes Imogen as having

———————" on her left breast
A mole cinque-spotted, like the crimson drops
I' the bottom of a cowslip."

But there is a variety with red flowers. They will sometimes flower again in November and December. Mr. Martyn speaks of some in his own gardens which always blew at that season when the winter was mild.

The light stalk of the Cowslip, gently bending with its weight of flowers, is elegantly described by Milton, who takes advantage of this drooping appearance to select it, with some others, to adorn the tomb of Lycidas:

" Bring the rathe-primrose that forsaken dies,
The tufted crow-toe, and pale jessamine,
The white pink, and the pansy freak'd with jet,
The glowing violet,
The musk-rose, and the well-attired woodbine,
With cowslips wan that hang the pensive head,
And every flower that sad embroidery wears:
Bid amaranthus all his beauty shed,
And daffodillies fill their cups with tears,
To strew the laureat hearse where Lycid lies."

And again, in the song of Sabrina, how beautifully does the unbending flower, and the airy tread of the goddess, each express the lightness of the other:

" By the rushy fringed bank,
Where grow the willow and the osier dank,

My sliding chariot stays,
Thick set with agate and the azure sheen
Of turkis blue, and emerald green,
That in the channel strays;
Whilst from off the waters fleet
Thus I set my printless feet,
O'er the cowslip's velvet head,
That bends not as I tread;
Gentle swain, at thy request,
I am here."

The oxlip is by no means so common as the Cowslip: it is considered as a link between that and the primrose. It has been called the great primrose: but though the oxlip flower spreads wider, the Cowslip has the advantage in height. On this account Shakspeare selects the latter for the courtiers of the Fairy Queen, in allusion to the tall military courtiers called Queen Elizabeth's Pensioners:

"The cowslips tall her pensioners be,
In their gold coats spots we see;
Those be rubies, fairy favours,
In those freckles live their savours;
I must go seek some dew-drops here,
And hang a pearl in every cowslip's ear."

The single Cowslip is rarely admitted into gardens, but the double flowers are common: they have a good effect by the side of the dark polyanthus, or shaded by a bunch of glowing wallflowers. The roots may be purchased almost for nothing. They who desire to have the single flowers may transplant the wild roots, which should be done about Michaelmas, and they will have time to gain strength for flowering in the spring. But it must be observed, that although these plants, in their wild state, are entrusted to Nature's care, and though we must confess that she deserves this confidence, we must no longer depend entirely upon her care of them, after we have removed them from her own great garden.

Cowslips love a moist soil; and when we plant them in a pot, the small portion of earth which it contains will naturally dry much faster than in the open ground: therefore, as we do not remove the brooks and springs with them, we must supply this deficiency by giving water to the potted plants in dry weather; in return for which, if we will find artists to manufacture it, they will furnish us with honey in abundance: for—

> "——— rich in vegetable gold
> From calyx pale the freckled cowslip born,
> Receives in amber cups the fragrant dews of morn."

COWSLIPS OF JERUSALEM.

PULMONARIA.

BORRAGINEÆ. PENTANDRIA MONOGYNIA.

Also called Sage of Jerusalem, Sage of Bethlehem, Spotted Comfrey, and Common Lungwort, as being esteemed in complaints of the lungs.—*French,* la grande pulmonaire; les herbes aux poumons; l'herbe du cœur [heart wort]; l'herbe au lait de Nôtre Dame [Our Lady's milk-wort]; pulmonaire d'Italie.—*Italian,* polmonaria maggiore.

This is a perennial plant, very much resembling the cowslip in form. The colours are many; not only on the same cluster, but even on the individual blossom, appearing various shades of red and blue, and these shades continually changing. Drayton places this flower in such honourable company, as gives us good reason to believe that he held it in great esteem:

> "Maids, get the choicest flowers, a garland, and entwine,
> Nor pinks, nor pansies, let there want; be sure of eglantine.

See that there be store of lilies,
(Called of shepherds daffodillies)
With roses damask, white, and red, the dearest flower-de-lis,
The cowslip of Jerusalem, and clove of Paradise."

DRAYTON'S PASTORALS.

CRINUM.

HEMEROCALLIDEÆ. HEXANDRIA MONOGYNIA.

THE Crinums most cultivated in this country are the American. The Great American Crinum flowers in July and August: the small species will flower three or four times in the year. They will thrive very well in a room generally inhabited in the winter; and their flowers at that time will be particularly valuable, so few being then in blossom. In the summer they should be placed abroad where they can enjoy the sunshine. The roots should be transplanted every year in March or October, and the offsets taken off and planted in separate pots, about six inches in diameter and eight or nine inches deep, filled with a light rich earth. Do not scruple to deprive the mother of her children, for she cannot afford food to so large a family; and the unnatural little bulbs will deprive her of all nourishment, and starve her without mercy, if they remain. The flowers are white, and sweet-scented. These plants should be watered *very sparingly* every second evening when newly planted; when they begin to shoot, they may have more water, *every* evening; but when they begin to blow, they will continue longer in blossom if more sparingly watered, as before.

CROCUS.

IRIDEÆ. TRIANDRIA MONOGYNIA.

An unhappy lover, whom the gods in pity were said to have changed into this flower.—*French,* safran.—*Italian,* zafferano; gruogo.

The Autumnal Crocus is supposed to have come originally from the East, but is now so common in Europe, that it is difficult to ascertain with certainty its original birth-place. The flowers are of a purple, lilac, or pale blue colour, blowing in October: the leaves grow all the winter. This species of Crocus is also called Saffron, and the medicine so called is obtained from it. Saffron was formerly more esteemed as a medicine than at present; but it is still used occasionally: it is often substituted for eggs in cakes, puddings, &c. and to some persons its flavour is very agreeable. A bag of saffron worn at the pit of the stomach has been lately said to be an effectual preventive of sea-sickness.

The first introduction of this plant into the country was considered so great a national benefit, as to have occasioned much controversy upon the subject. It is commonly said that Sir Thomas Smith was the first who brought it to England, in the reign of Edward the Third, and that it was first planted at Walden in Essex. That Walden was noted for the cultivation of it is clear, since the flower has even bestowed its own name upon that place, which is commonly called Saffron Walden. In Hakluyt's Voyages (edit. 1599, vol. ii. p. 165) the first introduction of Saffron is ascribed to a pilgrim, who, with the intention of serving his country, stole a head of Saffron, which he hid in his staff: but this is mentioned only as a thing reported at Saffron Walden*. Mr. Martyn, after referring to this volume, says he has been informed that the corporation of Walden bear three Saffron plants in their arms.

The Spring Crocus is common in many parts of Europe: there are many varieties; and as this kind furnishes the florists with seed, new varieties continually occur. The most usual are the Common Yellow, the Great Yellow, Deep Blue, Light Blue, White with Blue Stripes, Blue with White Stripes, White with a Purple Base, and Cream-coloured,—all natives of Britain: as also several from Scotland; the Black and White Striped, the Cloth of Gold, &c.

The Spring Crocus flowers in March; and where there are plenty of them, they make a magnificent show. If the season be mild, the flowers will sometimes appear in February, before the leaves have grown to any length. The leaves must not be cut off before they decay, or the root will be deprived of nourishment, and will not produce handsome flowers the next year. About the end of May, when the leaves and fibres have decayed, the roots may be taken up, wiped clean from earth, husk, &c. and placed in a dry room till September, when they should be replanted. Care must be taken to preserve them from mice, and other fond enemies: mice will utterly destroy them if they can get at them. The bulb should be planted with the bud uppermost, and the earth an inch deep

* This, however, is probably only a version of the history of the introduction of silk into Europe: two monks having brought from China, in the hollow of their walking-canes, the eggs of the silk-worm, which were hatched at Constantinople under the empress's own eye; who had, during the two years' absence of the monks, caused some mulberry trees to be got ready for the food of the young family.

above the top: for one root, a pot three inches wide will be large enough; four roots may be planted in a pot of six inches in diameter. They should be kept moist; which will require more or less water, according as they are in the sun or the shade, the room or the balcony, &c.: they will continue longer in blossom if watered rather sparingly after they have begun to blow. These bulbs will likewise flower in water: they may be put into the glasses any time from October to January, and thus be continued in succession. The water should rise a little above the widening of the glass; and from the time the fibres begin to shoot, should be renewed every four or five days.

The Autumnal Crocus does not increase so fast as the Spring kinds, nor does it produce seeds in this country. It should be replanted in August, as it flowers in September or October. These Crocuses will produce handsomer flowers if the bulbs be left undisturbed for two or three years; but they must not remain more than three. When it is not intended to remove them every year, more room must be allowed them for the growth of the offsets.

Virgil speaks of the Crocus as one of the flowers upon which bees love to feed:

———— "pascuntur et arbuta passim,
Et glaucas salices, casiamque, crocumque rubentem,
Et pinguem tiliam, et ferrugineos hyacinthos."
VIRGIL, GEORGIC 4.

"They feed also at large on arbutes and hoary willows, and cassia, and glowing saffron, and fat limes, and deep-coloured hyacinths."—MARTYN'S TRANSLATION, p. 372.

CYCLAMEN.

PRIMULACEÆ. PENTANDRIA MONOGYNIA.

This name is of Greek origin, and signifies circular. It alludes either to the roundness of the leaves, or of the roots. The familiar name among the country people is Sow-bread.—*French,* pain du porceau; in the village dialect, pan de pur, both signifying sow-bread. *Italian,* pane porcino; pane terreno [ground bread.]

THE common Cyclamen is an Austrian. The flowers are purple, drooping, and sweet-scented. The Ivy-leaved species is Italian: the flowers appear in August or September, soon after the leaves come out, and continue growing till May, when they begin to decay, and in June are quite dried up. There are two varieties; one with white, and one with purple flowers.

The Round-leaved Cyclamen is a native of the South of Europe: it has purple flowers, which blow late in the autumn.

The Persian Cyclamen, which is the most popular, flowers in March or April: it is sweet-scented, and varies in colour from a pure white to white and purple, or sometimes to a beautiful blush-colour. It is, as the name implies, a native of Persia: it has also been found in the Isle of Cyprus; and is, indeed, not unworthy of cultivation in Venus's own garden. It is a pretty flower for the parlour or study table; and the temperature of an inhabited room is well adapted to it.

The Cyclamen requires shelter from frost; particularly the two last-mentioned kinds. During the winter, or while destitute of leaves, they should have very little water, and be carefully preserved from damps. In the summer, they should be placed where they may enjoy the

I 2

sun till about eleven o'clock. They do not flower till the fifth year after they are sown.

CYTISUS.

LEGUMINOSÆ. DIADELPHIA DECANDRIA.

Said to have been first found in the island of Cythnus, whence it has derived its name.—*French,* le cytise.—*Italian,* citiso; avorniello; maggio pendolino.

Of this genus is that most elegant tree, the Laburnum, which drops its yellow blossoms so invitingly, as if wooing the beholder to pluck them. There are two varieties of Cytisus; one with narrower leaves and longer blossoms than the other, which is by far the handsomest, and is very justly called Golden-chain. I have seen a sprig of this in a lady's hair, where its bright green leaves, and its drooping blossoms, intermingling with the rich chestnut curls, had a very graceful appearance. But unfortunately it does not long survive the gathering: so that ladies who are disposed to adorn themselves with it must have recourse to imitation; and this, notwithstanding the perfection to which artificial flowers have been brought of late, will not easily equal the real flower.

It is well for the present purpose that the handsomest of the Laburnums is the smallest tree, and may be grown in a tub for many years. They ought to be in company with leafier trees, as they are but sparingly supplied with green of their own. Who would not have at least one of them, were it but to place by the side of the Persian lilac, or the rhododendron?

It has been recommended to sow the Laburnum in plantations infested with hares and rabbits; for so long as they can find a sprig of it, they will touch nothing else:

and though it be eaten to the ground in the winter, it will spring up again the next season, and thus be a constant supply for them. A whole plantation will be secured at the expense of a few shillings.

Laburnum-wood is very strong, and is much used for pegs, wedges, knife-handles, musical instruments, and a variety of purposes of that nature. Mr. Martyn, in his edition of Millar's Dictionary, speaks of a table and chairs made of this wood, which judges of elegant furniture pronounced to be the finest they had ever seen.

Pliny speaks of this wood as next in hardness to the ebony: it has been thought to make the best bows; and it occasionally afforded torches for the Roman sacrifices:

> "Tondentur cytisi; tædas sylva alta ministrat;
> Pascunturque ignes nocturni et lumina fundunt."
>
> VIRGIL, GEORGIC 2.

"The cytisus is cut, the tall wood affords torches, and the nocturnal fires are fed, and spread their light."—MARTYN'S TRANSLATION, p. 197.

The tree was formerly called Peas-cod-tree, and Bean-trefoil; but it is now generally known by its Latin name *Laburnum*, which is supposed to have been derived from the Alpine name *L'aubours*. The French call it *Cytise des Alps* (Cytisus of the Alps), and *Faux Ebenien* (False Ebony-tree). It is a native of Switzerland, Austria, and the Levant, &c. and flowers in May: at this season the mountains in Italy are hung so richly with its golden drapery as to obtain for it the name of Maggio, as we give that of May to the hawthorn.

The Black Cytisus is a shrub, seldom growing higher in this country than three or four feet: it is very bushy, and the branches are terminated by bunches of yellow flowers, four or five inches in length, having a very agreeable scent. It blossoms in July. This is a native of Silesia, Hungary, Italy, &c.

The Winged-leaved Cytisus is a handsome shrub, scarcely two feet high: the flowers are large, and of a deep yellow. It is a native of Siberia.

The Common Cytisus is a native of the South of Europe: it grows seven or eight feet high, is very bushy, and has bright yellow flowers. These will live all the year abroad after they are first raised to strength. If the weather be very dry, they should be watered once or twice a week.

Virgil recommends the Cytisus as a food for goats:

> "At cui lactis amor, cytisos, lotosque frequentes
> Ipse manu, salsaque ferat præsepibus herbas."
>
> VIRGIL, GEORGIC 3.

"Those who desire to have milk, must give them with their own hands plenty of cytisus and water-lilies, and lay salt herbs in their cribs."—MARTYN'S TRANSLATION, p. 313.

The bright blossoms of the Laburnum have not escaped the attention of our poets. Mr. Keats, in two distinct passages of his earliest poetry, each representing the flowery nook most beautiful to his fancy, gives a place to the Laburnum:

> "A bush of May-flowers with the bees about them;
> Ah, sure no tasteful nook would be without them:
> And let a lush laburnum oversweep them,
> And let long grass grow round the roots to keep them
> Moist, cool, and green; and shade the violets,
> That they may bind the moss in leafy nets.
>
> * * * * *
>
> Where the dark-leaved laburnum's drooping clusters
> Reflect athwart the stream their yellow lustres,
> And intertwined the cassia's arms unite
> With its own drooping buds, but very white."

> ——— "Laburnum, rich
> In streaming gold."
>
> COWPER'S TASK.

It is curious to observe how some plants appear to be compounded of others. Thus the Camellia Japonica has been noticed as resembling a bay-tree with roses; the arbutus is like another species of bay, yielding strawberries; and the Laburnum seems like a tree made up of large trefoil and garlands of yellow peas. The Geranium kind seems to delight in this species of mimicry.

When the Laburnum tree is so situated as to be shaded from the scorching suns of noon, it thrives so much better as to appear, to a superficial observer, a tree of a different kind.

DAHLIA.

CORYMBIFERÆ. SYNGENESIA POLYGAMIA SUPERFLUA.

THE Dahlia was named in honour of Andrew Dahl, a Swedish botanist. There are several species, all natives of the mountainous parts of the Spanish settlements in South America. The flowers are large and handsome; mostly red or purple, and the colours beautifully vivid. It is a very lofty plant, and the foliage is coarse and rank. It is thought to grow less luxuriantly, and to flower better, if planted in a poor and gravelly soil, in the open ground: they may, however, be obtained in pots. They will bear open air; and the roots will live a long time out of the earth without injury. The best time to plant them is in April. A recent improvement in the culture of this beautiful plant is to graft the young buds upon the tubers. They do not require much water.

This flower, comparatively a stranger in England till lately, from its great beauty has become very popular. When in flower, it makes a brilliant figure in the nursery-gardens, where many are planted together, and of various colours. It makes a fine show in a bouquet too, but will not long survive the gathering. The double flowers are as magnificent as the peony itself.

The best account of the Dahlia is to be found in the second part of the Transactions of the Horticultural Society, by R. A. Salisbury, Esq.

DAISY.

BELLIS.

CORYMBIFERÆ. SYNGENESIA POLYGAMIA SUPERFLUA.

The botanical name is derived from the Latin word *bellus*, handsome. In Yorkshire called Dog-daisy and Bairnwort. The word Daisy is a compound of day's and eye, Day's-eye; in which way, indeed, it is written by Ben Jonson.—*French*, la paquerette; paquerette vivace; paquette; marguerite [pearl]; petite marguerite; petite consire: in Languedoc, margarideta.—*Italian*, margheritena; margherita; pratellina [meadow-flower]; bellide; fiore di primavera [spring-tide-flower.]

Who can see, or hear the name of the Daisy, the common Field Daisy, without a thousand pleasurable associations! It is connected with the sports of childhood and with the pleasures of youth. We walk abroad to seek it; yet it is the very emblem of home. It is a favourite with man, woman, and child: it is the *robin* of flowers. Turn it all ways, and on every side you will find new beauty. You are attracted by the snowy white leaves, contrasted by the golden tuft in the centre, as it rears its head above the green grass: pluck it, and you will find it backed by a delicate star of green, and tipped with a blush-colour, or a bright crimson.

"Daisies with their pinky lashes"

are among the first darlings of spring. They are in flower

almost all the year; closing in the evening and in wet weather, and opening on the return of the sun:

> "The little dazie, that at evening closes."
> SPENSER.

> "By a daisy, whose leaves spread
> Shut when Titan goes to bed."
> G. WITHERS.

No flower has been more frequently celebrated by our poets, our best poets; Chaucer, in particular, expatiates at great length upon it. He tells us that the Queen Alceste, who sacrificed her own life to save that of her husband Admetus, and who was afterwards restored to the world by Hercules, was, for her great goodness, changed into a Daisy. He is never weary of praising this little flower:

> ——————" Whan that the month of May
> Is comen, and that I heare the foules sing,
> And that the floures ginnen for to spring,
> Farewell my booke, and my devocion,
> Now have I than eke this condicion,
> That of all the floures in the mede,
> Than love I most these floures white and rede,
> Such that men callen daisies in our town:
> To them I have so great affectioun,
> As I sayd erst, whan comen in the Maie,
> That in my bedde there daweth me no daie,
> That I nam up, and walking in the mede
> To seen this floure ayenst the sunne sprede,
> Whan it upriseth early by the morrow,
> That blissful sight softeneth my sorow,
> So glad am I, when that I have presence
> Of it, to done it all reverence,
> As she that is of all floures the floure,
> Fulfilled of all vertue and honoure,
> And every ilike faire, and fresh of hewe,
> And ever I love it, and ever ilike newe,
> And ever shall, until mine herte die,
> All sweare I not, of this I woll not lie.
> There loved no wight nothen in this life,
> And whan that it is eve I renne blithe,

As soone as ever the sunne ginneth west,
To seen this floure, how it woll go to rest,
For feare of night, so hateth she darkenesse,
Her chere is plainly spred in the brightnesse
Of the sunne, for there it woll unclose:

* * * * *

My busie ghost, that thursteth alway new,
To seen this floure so yong, so fresh of hew,
Constrained me with so gredy desire,
That in my haste, I fele yet the fire,
That made me rise ere it were day
And this was now the first morowe of Maie,
With dreadfull herte, and glad devocion
For to been at the resurrection
Of this floure, whan that it should unclose.
Again the sunne, that rose as redde as rose,
That in the brest was of the beast that day
That Angenores daughter ladde away.
And doune on knees anon right I me sette,
And as I coulde, this fresh floure I grette,
Kneeling alway till it unclosed was,
Upon the small soft swete grass,
That was with floures swete embrouded all,
Of such sweteness, and odour over all,
That for to speak of gomme, herbe, or tree,
Comparison may not imaked be,
For it surmounteth plainly all odoures,
And of riche beaute of floures.

* * * * *

And Zephyrus and Flora gentelly
Yave to the floures soft and tenderly,
Hir swete breth, and made hem for to sprede,
As god and goddesse of the flourie mede,
In which me thought I might day by daie,
Dwellen alway the joly month of Maie,
Withouten slepe, withouten meat, or drinke:
Adowne full softly I gan to sinke,
And leaning on my elbow and my side,
The long day I shope me for to abide,
For nothing els and I shall not lie,
But for to look upon the daisie,
That well by reason men it call may
The daisie, or els the iye of the day,

The emprise, and floure of floures all,
I pray to God, that faire mote she fall,
And all that loven floures for her sake:
* * * * *
And from a ferre come walking in the mede,
The god of love, and in his hand a queene,
And she was clad in royal habit greene,
A fret of golde she had next her heere,
And upon that a white croune she bare,
With narrow small, and I shall not lie,
For all the world right as a daisie
Icrouned is, with white leaves lite,
So were the florounes of her croune white,
And of a perle fine orientall,
Her white croune was imaked all,
For which the white croune above the grene
Made her like a daisie for to seme,
Considred eke her fret of gold above:
* * * * *
Quod Love * * * *
* * * * *
Hast thou not a book in thy cheste
The great goodnesse of the Queene Alceste
That turned was into a daisie,
She that for her husband chose to die,
And eke to gone to hell rather than he,
And Hercules rescued her parde
And brought her out of hell again to bliss?
And I answerde againe, and said, 'Yes,
Now I knowe her, and is this good Alceste,
The daisie, and mine owne hertes rest?'" *

Chaucer makes a perfect plaything of the Daisy. Not contented with calling to our minds its etymology as the eye of day, he seems to delight in twisting it into every possible form; and, by some name or other, introduces it continually. Commending the showers of April, as bringing forward the May flowers, he adds:

"And in speciall one called se of the daie,
The daisie, a flower white and rede,

* See Chaucer's Prologue to the Legend of Good Women.

And in Frenche called La Bel Margarete.
O commendable floure, and most in minde!
O floure and gracious of excellence!
O amiable Margarite! of natife kind"—

In another poem, describing an arbour, he says:

"With margarettes growing in ordinaunce
To shewe hem selfe as folke went to and fro,
That to beholde it was a great plesaunce,
And how they were accompanied with mo,
Ne momblisnesse and soneness also
The poure pensis were not dislogid there,
Ne God wote ther place was every where."

He tells us that the Queen Alceste was changed into this flower: that she had as many virtues as there are florets in it.

"Cybilla made the daisie, and the flour
Icrownid all with white, as man may se,
And Mars yave her a corown red, parde,
In stede of rubies set among the white."

"The daisy scattered on each meade and downe,
A golden tufte within a silver croune.
Fayre fall that dainty flowre! and may there be
No shepherd graced that doth not honor thee!"
W. Browne.

But the Field Daisy is not an inhabitant of the flower-garden: it were vain to cultivate it there. We have but to walk into the fields, and there is a profusion for us. It is the favourite of the great garden of Nature:

"Meadows trim with daisies pied."

The reader will doubtless remember Burns's Address to a Mountain Daisy, beginning

"Wee, modest, crimson-tipped flower."

The Scotch commonly call it by the name of Gowan; a

name which they likewise apply to the dandelion, hawk-weed, &c.:

"The opening gowan, wet with dew."

Wordsworth, with a true poet's delight in the simplest beauties of nature, has addressed several little poems to the Daisy:

"In youth from rock to rock I went,
From hill to hill, in discontent
Of pleasure high and turbulent,
  Most pleased when most uneasy;
But now my own delights I make,—
My thirst at every rill can slake,
And gladly Nature's love partake
  Of thee, sweet daisy!

"When soothed awhile by milder airs,
Thee Winter in the garland wears
That thinly shades his few grey hairs;
  Spring cannot shun thee;
Whole summer fields are thine by right;
And Autumn, melancholy wight,
Doth in thy crimson head delight
  When rains are on thee.

"In shoals and bands, a morrice train,
Thou greet'st the traveller in the lane;
If welcomed once, thou count'st it gain;
  Thou art not daunted,
Nor carest if thou be set at nought:
And oft alone in nooks remote
We meet thee, like a pleasant thought,
  When such are wanted.

"Be violets in their secret mews
The flowers the wanton Zephyrs choose;
Proud be the rose, with rains and dews
  Her head impearling;
Thou liv'st with less ambitious aim,
Yet hast not gone without thy fame;
Thou art indeed by many a claim
  The poet's darling.

"If to a rock from rains he fly,
Or some bright day of April sky,
Imprisoned by hot sunshine lie
Near the green holly,
And wearily at length should fare;
He need but look about, and there
Thou art!—a friend at hand, to scare
His melancholy.

"A hundred times, by rock or bower,
Ere thus I have lain couched an hour,
Have I derived from thy sweet power
Some apprehension;
Some steady love; some brief delight;
Some memory that had taken flight;
Some chime of fancy, wrong or right;
Or stray invention.

"If stately passions in me burn,
And one chance look to thee should turn,
I drink out of an humbler urn
A lowlier pleasure;
The homely sympathy that heeds
The common life, our nature breeds;
A wisdom fitted to the needs
Of hearts at leisure.

"When, smitten by the morning ray,
I see thee rise alert and gay,
Then, cheerful flower! my spirits play
With kindred gladness:
And when, at dusk, by dews opprest
Thou sink'st, the image of thy rest
Hath often eased my pensive breast
Of careful sadness.

"And all day long I number yet,
All seasons through, another debt,
Which I, wherever thou art met,
To thee am owing;
An instinct call it, a blind sense;
A happy genial influence,
Coming one knows not how nor whence,
Nor whither going.

"Child of the Year! that round dost run
Thy course, bold lover of the sun,
And cheerful when the day 's begun
As morning leveret,
Thy * long-lost praise thou shalt regain;
Dear shalt thou be to future men
As in old time;—thou, not in vain,
Art Nature's favourite."

Nor in vain is it a favourite with the poet, who emulates Chaucer himself in doing it honour. At one time he describes it as

"A nun demure, of lowly port;
Or sprightly maiden of Love's court,
In her simplicity the sport
Of all temptations.
A queen in crown of rubies drest;
A starveling in a scanty vest;
Are all as seems to suit *it* best,
Its appellations.

"A little Cyclops with one eye
Staring to threaten and defy,
That thought comes next,—and instantly
The freak is over,
The shape will vanish; and, behold!
A silver shield with boss of gold,
That spreads itself, some faery bold
In fight to cover."

But again we must remember this is not to be a reprint of Mr. Wordsworth's poems.

Of the Garden Daisy there are many varieties: the Double White; Red; Red and White Striped; the Variegated; the Proliferous, or Hen and Chicken, &c. These, indeed, are but double varieties of the Field Daisy, but less prolific, and flowering only for a few months—April, May, and June.

* See in Chaucer and the elder poets, the honours formerly paid to this flower.

The Annual resembles the Common Daisy, but is not so large: it is a native of Sicily, Spain, Montpelier, Verona, and Nice.

The Garden Daisy should be planted in a loamy, unmanured earth, and placed in the shade; as the full noonday sun will sometimes kill it. The roots should be parted every autumn: they should be taken up in September or October, parted into single plants, and put in pots about five inches wide. When in pots, they will require a little water every evening in dry weather.

Rousseau, in his Letters on Botany, gives a long and beautiful description of the structure of the Daisy.

DANEWORT.

SAMBUCUS EBULUS.

CAPRIFOLIEÆ. PENTANDRIA TRIGYNIA.

Dwarf Elder, Wallwort, and Walewort.—*French,* yeble; hièble; petite sureau: in Provence, saupuden.—*Italian,* ebbio; ebulo.

DANEWORT is a shrub which grows three or four feet high, and bears a profusion of blossoms, of a dull red colour. It is a native of England, and many other parts of Europe; and was named Danewort among us from a notion that it had first sprung from the blood of the Danes. It blows in July, is very hardy, and likes a moist soil. Its leaves, like those of the common elder, are strewed to keep away moles and mice, which will not come near them. The elder tree is supposed to be prejudicial to persons reclining under its shade.

DAPHNE.

THYMELEÆ. OCTANDRIA MONOGYNIA.

This genus is named from the nymph beloved of Apollo: some of the species greatly resembling the bay.

Of this genus, the most beautiful kind, and the kind most frequent in our gardens, is the Daphne Mezereon: also called Spurge-olive, German Olive-spurge, Spurge-flax, Flowering-spurge, and Dwarf Bay. Most of the European languages give it a name equivalent to Female Bay. The French call it *laureole femelle; laureole gentille; bois joli; bois gentille; mal-herbe:* in the villages, *dzentelliet.*—The Italians, *Daphnoide; laureola femina; biondella* [little fair-one]; *camelea,* and *calmolea.*

The Daphne Mezereon is a handsome shrub: the flowers come out before the leaves, early in the spring; they grow in clusters all round the shoots of the former year. Thus it is as Cowper says:

> "Though leafless well attired, and thick beset
> With blushing wreaths, investing every spray."

It is a native of almost every part of Europe: with us, it is very common in the beech woods in Buckinghamshire. The name Mezereon is said to have been borrowed from the Dutch.

The branches of the Daphne Mezereon make a good yellow dye. The berries are a powerful poison, but the bark is a very useful and valuable medicine. The two principal varieties of this species of the Daphne are the White-flowered, which has yellow berries, and the Peach-coloured, of which the berries are red.

The Mezereon is very sweet-scented; and, where there

are many together, they will perfume the air to a considerable distance. The best time for transplanting this shrub is the autumn; because, as it begins to vegetate early in the spring, it should not be then disturbed. It thrives best in a dry soil: if it has too much wet, it becomes mossy, and stinted in its growth, and produces fewer flowers. It should enjoy the morning sun, and remain abroad all the year.

Of the other species of Daphne, the Silvery-leaved, a native of the South of France, is one of the prettiest. This will not bear transplanting, and must be sheltered in severe frost. It should be sparingly watered. The leaves of this shrub are white, small, soft, and shining like satin: between these leaves come out thick clusters of white flowers, bell-shaped, and tinged with yellow on the inside.

The Trailing Daphne grows naturally in many parts of Europe: it is remarkably sweet-scented, and has purple or white flowers, which appear very early in spring. It may be treated like the Mezereon, but that it will not bear transplanting.

The Spurge-laurel, *Daphne laureola*—in French, *laureole mâle; laureole des Anglois:* in Italian, *laureola maschio*—has flowers of a yellowish green, which, if the season be not very severe, come out soon after Christmas. It is a native of Britain, and many other parts of Europe. This shrub, like the Mezereon, is very useful as a medicine; but, like that also, should be trusted to the skill of experienced persons only. It is a hardy plant, and may be treated like the Mezereon. The plants may be removed from the woods, or elsewhere, in the autumn; and at that season may be increased by cuttings.

The Alpine may be treated like the Trailing Daphne. The Flax-leaved Daphne has flowers like the Mezereon, only smaller: they blow in June, and are very sweet-scented. This is a native of Spain, Italy, and the South of France: it will sometimes flower twice in the year. Gerarde gives this shrub the name of Mountain Widow-wayle. It may be treated in the same manner as the Silvery-leaved.

An excellent writing-paper is made from a tree of this genus, called the Daphne Cannabina, a native of Cochin-China. There is another, called the Lace-bark Daphne, of which the inner bark is of such a texture, that it may be drawn out in long webs like lace, and has been actually worn as such. Charles the Second had a cravat made of it, which was presented to him by Sir Thomas Lynch when governor of Jamaica. It is there principally used for ropes. This tree is a native of Jamaica, where it is called *Lagetto*, or Lace-bark-tree; and of Hispaniola, where it is known by the name of *Bois-dentelle* [Lace-wood]. It will not thrive in England, except in a stove. But this latter plant is now removed by Lamarcke, and forms a separate genus, *Lagetta*, which is placed by him, even in another family, the *Eleagneæ*.

DOG'S-BANE.

APOCYNUM.

APOCINEÆ. PENTANDRIA DIGYNIA.

So named from a notion that it is fatal to dogs.

THE Tutsan-leaved Dog's-bane is an extremely curious plant: the flowers are white, or pale red; bell-shaped, and the anthers are so constituted within it, as to entangle the flies who are attracted by the honey-juice it contains: so that in August, when in full flower, it is usually found full of their dead bodies. The French, in Canada, call it *Herbe-à-la-puce* [Fleawort], and say it is noxious to some persons, though harmless to others. Mr. Martyn quotes an author (Kalm) who mentions having seen a soldier whose hands were blistered all over merely from plucking it; whereas he frequently rubbed his own hands with the juice without feeling any inconvenience.

Mr. Lambert, in his Travels in Canada and the United States, affirms that he has seen several persons who have been confined to the house in consequence of having been poisoned in the woods by this plant, and that even the merely treading on it is sufficient to create swellings and inflammations: "and yet," continues he, "I have seen other people handle it with safety; and have myself often pulled it up by the root, broke the stem, and covered my hands with the milky juice which it contains, without experiencing any disagreeable effect. What property it is in the constitution of people which thus imbibes or repels the poisonous qualities of this plant, I have never been able to learn, nor can I from observation account for it.

"Many gardens in Lower Canada are full of Dog's-bane, which occasions it to be considered there as a weed. The roots appear to spread under ground to a considerable extent; and though the plant may be cut off every year, it springs up again in another place. It makes its appearance about the end of May, and runs up like the scarlet-beans, entwining itself round any tree, plant, or paling that stands in its way; and if there is nothing else upon which the young shoots can support themselves, they adhere to each other. Their leaves and stems are of a light green, and they are in full flower in July. Wherever the *Herbe-à-la-puce* grows, there is always to be found a great number of lady-flies [coccinella]. They are covered with a brilliant gold as long as they are on the leaf, or retain any particle of its juice. I caught some of them, and put them into a phial; but neglecting to add some leaves of the *Herbe-à-la-puce*, they had by the next morning lost their splendid coat, and merely resembled the common red lady-fly which we have in England. I then caught a few more, and having supplied them well with the leaves of that plant, they retained their gold tinge equally as well as in the open air. In a few days they had reduced the leaves to mere skeletons, but as long as there remained a morsel of the stalks or fibres to feed upon, their beautiful appearance continued. I kept them upwards of a month in this manner; giving them occasionally fresh leaves of the plant, and admitting the air through some holes that I pricked in the paper with which I had covered the mouth of the phial. They would feed upon no other plant than the *Herbe-à-la-puce*, from which alone they derived their beauty. I afterwards gave them their liberty, and they flew away apparently little the worse for their confinement*."

The Tutsan-leaved, the St. John's-wort-leaved, and the Spear-leaved kinds will bear the open air, if not exposed to too much wet; which, as they are very succulent, would rot them. They may be increased by parting the roots, which should be done in March.

Hemp Dog's-bane, a native of North America, is used by the Indians for various purposes: they prepare the stalks as we do hemp, and make twine, fishing-nets and lines, bags, and linen of them. According to Kalm, this is the species which the Canadians call *Herbe-à-la-puce*.

In Mrs. Charlotte Smith's Conversations are some lines upon the fate of a poor fly, lured to its prison by the deceitful sweetness of the Apocynum.

* Lambert's Travels through Canada, &c. vol. i. p. 435.

DRAGON'S-HEAD.

DRACOCEPHALUM.

LABIATÆ. DIDYNAMIA GYMNOSPERMIA.

So named from the form of the flower, which resembles a gaping mouth.

The Moldavian Dragon's-head, commonly called Moldavian Balm—in French, *la melisse de Moldavie; la Moldavique; la melisse des Turcs* [Turk's balm]—is an annual plant with blue flowers, which appear in July, and continue till the middle of August. It has a strong scent, which to some persons is very agreeable. The seeds of this, and of the other annual kinds, may be sown either in March or September, in small pots, one seed in each, or several in a larger pot, and thinned as they may require it when they come up. Some kinds are more branched than others.

Balm of Gilead is a perennial plant, a native of the Canary Islands: it has blue or flesh-coloured flowers, continuing from July to September. It is called Balm of Gilead from its fine odour when rubbed. The old writers call it Camphorosma, and Cedronella, upon the same account. It should be sown in September, and kept in the house during the winter. When grown, it will require shelter from frost only; but when first sown, should be treated rather more tenderly. It may also be increased by cuttings, which, planted in any of the summer months, and placed in the shade, will soon take root.

The Virginian Dragon's-head—named by the French, *la cataleptique; l'herbe aux paralitiques*, from its use in palsy and similar diseases—is a native of North America: it has purple flowers, blowing from July to September. This species requires a moist soil, and should have more water than the others. It may be increased by parting the roots in autumn.

The Austrian species is very handsome: the flowers are violet-coloured. The Hyssop-leaved, a native of Norway, Siberia, &c. has blue flowers, blowing in June.

The Siberian kind has pale blue flowers. The three last-mentioned species should be sown towards the end of March: when about two inches high, they should be removed from where they were first sown into separate pots, about eight inches wide: this should be done carefully, without removing the ball of earth attached to them, and they should be placed in the shade until they have taken new root. They will last three or four years, sometimes longer; but will not flower well after that age. They should all be kept tolerably moist, particularly when newly planted. The Balm of Gilead is the only kind that requires shelter in the winter.

DRYAS.

ROSACEÆ. ICOSANDRIA POLYGYNIA.

So called by Linnæus from the Dryades, or nymphs of the oaks: the leaves bearing some resemblance to those of the oak.

The Five-petaled Dryas is a native of Siberia: it is a very small plant, bearing yellow flowers. The Eight-petaled species is a delicate little evergreen, with snow-white blossoms, and extremely pretty: it is a native of Lapland, Denmark, Siberia, Ireland, Scotland, England, Italy, &c. Thus, like Homer, it may boast that at least seven different places claim the honour of its birth. It is a perennial plant, and will not suffer from cold. The earth should be kept moderately moist. It flowers in June.

EGG-PLANT.

SOLANUM MELONGENA.

SOLANEÆ. PENTANDRIA MONOGYNIA.

The inhabitants of the British islands in the West Indies call it Brown-John, or Brown-Jolly —*French,* mayenne; aubergine; beringene; verangeane; plante à œuf.—*Italian,* melanzana [mad apple]; uovo Turco [Turkish egg]; petronciano; marignano.

All the varieties of the Egg-plant are annual, and must be raised in a hot-bed: they are cultivated chiefly for their fruit, which is formed like an egg, and when white, has exactly the appearance of one: it varies in size from two to nine or ten inches in length; and in colour, from white to yellow, pale red, or purple. Here the fruit is only regarded as a curiosity; but in the East Indies they broil it, and eat it with salt and pepper; or slice it, pickle it for an hour or two, boil it tender, and eat it as greens. The Turks, who are fond of it, call it Badinjan.

It may be placed in the open air at the end of May. The fruit appears in July; and then, when the weather is dry, water should be given liberally every evening.

ERINUS.

RHINANTHACEÆ. DIDYNAMIA ANGIOSPERMIA.

French, l'erine; la mandeline.

The Alpine Erinus is a pretty little plant, producing flowers of a lively purple, which are in bloom the greater part of the summer; appearing in April or May. It is a native of Germany, the Swiss Alps, the Pyrenees, and the South of France. It must not be set in a rich soil: it prefers a loamy earth, is fond of the shade, and may be increased by parting the roots in autumn. The Erinus grows naturally among the rocks, and will thrive well in the chinks of an old wall: it should have a little water in dry weather.

EVERLASTING.

GNAPHALIUM.

CORYMBIFERÆ. SYNGENESIA POLYGAMIA SUPERFLUA.

The botanical name is derived from the Greek, and signifies cotton, or nap: the origin of the familiar name is obvious.—*French,* gnaphale; cottoniere.—*Italian,* gnafalio; elicriso.

The Common Shrubby Everlasting grows to the height of about three feet. It has yellow flowers, which, if gathered before they are much opened, and kept from air and dust, will continue in beauty many years. It is a native of Germany, France, and Spain. By old writers, it is called Gold-flower, God's-flower, Goldilocks, Golden-stoechas, and Cassidony. This species, the Red-flowered, and the Sweet-scented with yellow flowers, are sufficiently hardy to live in the open air in mild winters, if placed in a sheltered situation; but it is advisable to cover their roots with straw in frosty weather; and if very severe, they must be housed. The French usually call it *l' immortelle jaune;* but in Languedoc, *sauveto.*

The Pearly-White Everlasting is a native of North America, where it is called Life-Everlasting, because its silvery leaves will long preserve their beauty unchanged. It grows in extreme profusion in uncultivated fields, glades, &c. and flowers from July to September. A decoction of the stalks and flowers is used to foment the limbs for pains and bruises. The Plantain-leaved is also a North American, and has white flowers. These two kinds will thrive in almost any soil or situation, and are easily increased by their roots, which may be transplanted in the autumn.

There are many species from the Cape, requiring winter shelter, but not artificial heat: their flowers are white, purple, or yellow. The earth should be kept moderately moist for all the species.

The Eastern-Everlasting, called Golden-Flower-Gentle, is one of the Cape kinds: it has been long cultivated in Portugal, where, in the winter season, the churches are adorned with its brilliant flowers.

FOX-GLOVE.

DIGITALIS.

PERSONEÆ. DIDYNAMIA ANGIOSPERMIA.

This plant is also called Finger-Flower; the shape of the flower resembling the finger of a glove; and Bell-Flower.—*French,* dogtier [finger-flower], gantlet; gants de notre dame.—*Italian,* guantelli; aralda.

The common Fox-glove is an extremely handsome flower, varying in colour from a Roman-purple to a violet-colour, cream-colour, orange-tawney, blush-colour, or white. It has a poisonous quality, but in skilful hands becomes a useful medicine. This species is a native of Denmark, Germany, Switzerland, and Great Britain; and flowers from June to August or September.

The Iron-coloured Fox-glove is a native of Italy and Constantinople, and flowers from the beginning of June to the end of July.

The seeds should be sown in autumn, about four in a pot seven or eight inches wide: in dry summer weather they should be watered every evening; but in the winter two or three times a week would be sufficient. The Canary Shrubby Fox-glove has yellow flowers, which begin to appear in May. The Madeira-Shrubby Fox-glove is a very handsome plant, flowering in July and August. These two kinds must be sheltered in the winter, admitting the fresh air in mild weather. In the summer they should be placed abroad where they may enjoy the morning sun; and in dry weather be plentifully watered every evening: twice a week will suffice in winter.

The seedlings, sown in the autumn, and kept in a room tolerably warm, will be large enough to be transplanted early in spring; when they should be placed separately in small pots, and after they have taken firm root, should be gradually accustomed to the open air.

It is a pity this plant is poisonous, for it is extremely beautiful, particularly those kinds which are of a deep rose-colour. They are all speckled within the bell, which adds still more to their richness.

Mrs. C. Smith invites the bee to

> "Explore the Fox-glove's freckled bell."

Browne uses a similar epithet when he describes Pan as seeking gloves for his mistress; a curious conceit:

> "To keepe her slender fingers from the sunne,
> Pan through the pastures oftentimes hath runne
> To pluck the speckled Fox-gloves from their stem,
> And on those fingers neatly placed them."
>
> W. BROWNE.

It is not one of his happiest passages; but he is a true poet, and deserves in particular the gratitude of the lover of nature. Cowley has the same conceit, but conceits are common with Cowley:

> "The Fox-glove on fair Flora's hand is worn,
> Lest while she gathers flowers she meet a thorn."
>
> COWLEY, ON PLANTS.

FRITILLARY.

FRITILLARIA.

LILIACEÆ. HEXANDRIA MONOGYNIA.

The Imperial Fritillary, or Crown Imperial, is supposed to be a native of Persia: there are many varieties, all handsome, and varying in colour. This species is less esteemed than its beauty merits, on account of its strong and disagreeable scent. The earth should be kept moderately moist. About the end of July the bulbs should be taken up, cleaned, &c. &c. and kept out of the earth about two months; but care must be taken in putting them by, to lay them all singly, not in heaps. The offsets should be the first planted, because they are the most apt to shrink.

This lily requires deep pots, and the bulb should be laid four inches deep at the least. It will require support, and will flower in April.

The Persian Fritillary, or Persian Lily—called by the Italians, *giglio di Persia; giglio di Susa;* and *pennacchi Persiani*—bears a spike of deep purple flowers, growing at the top of the stem in the form of a pyramid: they open in May, but seldom produce seeds in England.

The Black Fritillary is a native of France and Russia; it has yellow flowers, which blow in April or May.

The Common Fritillary, or Chequered Lily, is a native of England, and most of the southern parts of Europe. The flowers are chequered with purple and white, or purple and yellow. "It is for this reason," says Mr. Martyn, "that it has been named Fritillaria, from *fritillus,* [a draught or chess-board]. Nevertheless, fritillus is *not* the board, but the dice-box."

It has many familiar English names, as Turkey-hen-flower, Guinea-hen-flower, Chequered Daffodil, and Snake's-head; from which last name, a meadow between Kew and Mortlake is called Snake's-head Meadow. Some call it Narcissus Caparonius, from Noel Caparon, who first discovered it: he was an apothecary, then dwelling in Orleans, but murdered soon after in the massacre of France. The French call it *le damier* [the chess-board]: the Italians, *giglio variegato; fritillaria scaccheggiata* [chess-board fritillary].

Gerarde informs us that "the curious and painful herbalist of Paris, John Robin," sent him many of the plants for his garden, and that "they were greatly esteemed for the beautifying of our gardens, and the bosoms of the beautiful."

This lily flowers in April and May. The three last may be preserved and increased in the same manner as the first; only they do not take so much room, and the roots will not keep so long out of the ground. If they cannot conveniently be planted sooner, they should be laid in sand, to prevent their shrinking. They should be removed every second year; will bear the open air; and should have just water enough to prevent drought.

FUCHSIA.

MYRTOIDEÆ. TETRANDRIA MONOGYNIA.

So named in honour of Leonard Fuchs, a noted German botanist.

THIS is a most beautiful little plant; the leaves are of a fine green; their veins tinged with red: the flowers pendulous, and of a brilliant scarlet. "The Scarlet Fuchsia," says Mr. Martyn, "is a plant of peculiar beauty, producing its rich pendant blossoms during most part of the summer: the petals in the centre of the flower are particularly deserving of notice; they somewhat resemble a small roll of the richest purple-coloured riband."

It is a native of Chili. It will not bear the open air in this country, and in the winter must be kept in a warm inhabited room; for it is commonly treated as a stove-plant. The fresh air should be admitted in the summer, and it should always be kept moist. This is an elegant plant for the drawing-room or study.

FUMITORY.

FUMARIA.

FUMARIDEÆ. DIADELPHIA HEXANDRIA.

So named from a notion that it affects the eyes like smoke, or rather because a bed of the common kind when in flower appears at a distance like a dense smoke.—*French,* fumiterre; the common species by the villagers is called coridalo.

THE Red Canadian Fumitory is a handsome plant with large flowers. The Evergreen Fumitory has purple flowers, which bloom all the summer: it is a native of North America. They may be sown in the autumn; two or three seeds in a pot eight inches wide; and watered occasionally in dry weather. The Naked-stalked Fumitory, a native of Canada, has white and yellow blossoms, and may be increased by offsets from the roots, which should be planted in the autumn when the leaves have decayed.

The Great-flowered Siberian, with white and yellow flowers; blowing in May: and the Bulbous Fumitory, with purple, blush-coloured, or white flowers, may also be increased by offsets. The time to transplant these is between May and August (inclusive), as the leaves die off.

They are pretty, and very hardy. As they do not increase very fast, they should not be parted oftener than once in three years. They like a light sandy soil, and the earth should be kept moderately moist. The last species grows wild in many parts of Europe. It has the scent of the Cowslip.

The Bladdered Fumitory is rather tender. It is an annual plant, raised in a hot-bed, and not exposed to the open air till June. The other kinds are hardy, and may be treated in the same manner as those first mentioned.

GENTIAN.

GENTIANA.

GENTIANEÆ. PENTANDRIA DIGYNIA.

So named from Gentius, King of Illyria.—*French,* la gentiane.—*Italian,* la genziana.

The Gentians are very numerous, and many of them eminently beautiful. They are generally difficult to preserve in a garden; and being long-rooted, very few are adapted for planting in pots. The smaller kinds, however, may be so cultivated; as the Swallow-wort-leaved, which does not exceed a foot in height, and has large light-blue bell-shaped flowers, blowing in July and August. This species is a native of many parts of Europe; it must have a moist loamy soil, and be placed in the shade. The roots only are perennial; the stalks decay annually; and of most of the species, the flowers appear but once in two or three years. They all like moisture, and should be watered liberally in dry weather, particularly the March Gentian, which has also fine blue flowers, though few in number. They blow in August and September. This species grows naturally in England and many other parts of Europe. If it be in strong moist earth, it will flower every year.

Cowley, taking advantage of the origin of the name, proposes this plant, by the name of Royal Gentian, in answer to Virgil:

> "Dic quibus in terris inscripti nomina regum
> Nascantur flores."
>
> VIRGIL, Eclogue 3.

> "Now tell me first in what new region springs
> A flower that bears inscribed the name of kings."
>
> DRYDEN.

Not very aptly, however, for the Gentian is not supposed to bear a name inscribed on its blossoms. Virgil has been supposed to allude to the Hyacinth and some of the poetical fictions connected with it.

The Dwarf Gentian, or Gentianella, has a most beautiful blue flower, which blows in April and May: it is a native of the Alps, requires the same treatment as the others, and may be increased by parting the roots once in three or four years. The Fringed-flowered species has also large blue flowers, appearing in August and September. It is a native of many parts of Europe, and of Canada, and may be treated in every respect like the last.

The Small Alpine, and the aquatic kinds have also blossoms of a vivid blue, flowering in May and June. Linnæus speaks of the first as adorning the Pyrenees with its splendid blossoms; the latter is a native of China and Japan. These two are annual, and, growing naturally in wet spongy places, should be sown in a boggy earth, and placed in the shade. When the plants come up, the surface of the earth should be covered with moss, which should be kept always moist, or rather—wet. Of most of the species, there is a variety with white flowers.

The Large Yellow Gentian is a very useful plant, being

not only a valuable medicine, but also an excellent substitute for hops in brewing: and before hops had established their reputation, this Gentian was commonly used for that purpose. The roots of this and of the Purple Gentian strike two feet in depth.

All the kinds here mentioned will bear the open air.

GERANIUM.

GERANIUM, ERODIUM, AND PELARGONIUM.

GERANIACEÆ. MONADELPHIA DECANDRIA

The name, Geranium, is derived from the Greek language, and signifies a crane: the fruit having the form of a crane's bill and head. The English name is Crane's-bill; but the plant is more generally known by its botanical appellations. The Geranium is divided into three genera: Erodium is the first, Pelargonium the second, and the third retains the old name of Geranium which, indeed, is still familiarly used for them all, as well as the English name Crane's-bill. Erodium is from the Greek, and signifies a heron, whose bill is similar to that of the crane; Pelargonium is from the same language, and signifies a stork, whose bill is equally long.—*French,* le geranion; la geraine; bec de grue; bec de cicogne.—*Italian,* geranio, becco di gru.

THERE is no end to the varieties of Geranium, and as new ones continually occur, there most probably never will be an end to them. It were idle to attempt a general description of a plant so well known as any common species of Geranium; since there is scarcely a street, or even an alley in London, but is adorned with one or more of them. But there are many plants bearing this title which have no kind of resemblance to these in their general appearance, and which the most passionate lover or attentive observer of these beautiful plants, unskilled in the mysteries of botanical science, would never discover to belong to them.

The Erodiums, with very few exceptions, may be increased—the annual kinds from seed, the perennial by parting the roots in autumn,—and will thrive in the open air. The principal exceptions are the Crassifolium, or Upright Crane's-bill, the Incarnatum, or Flesh-coloured, the Glaucophyllum, or Glaucous-leaved, and Chamædryoides, or Dwarf Geranium, which must be treated as the Pelargoniums.

The Geranium, specifically so called, may be treated in the same manner as the Erodiums, and will thrive in almost any soil or situation. The Pelargoniums, which constitute the principal division of this great genus, require more care. They may be easily raised from seed; but a person desiring large and early flowers will procure a plant which has been raised in a hot-bed.

The Shrubby African Geraniums are commonly increased by cuttings, which, planted in June or July, and placed in the shade, will take root in five or six weeks. In September, or in October, as the weather is more or less mild, they must be housed: even when grown, the Pelargoniums must be housed in winter; at which time they should be gently watered twice a week, if the weather is not frosty. In May they may be gradually accustomed to the open air, and about the end of that month be placed abroad entirely in the day; but should still for the next two or three weeks be under cover at night, though fresh air must be admitted. After that time they must be defended from strong winds, and be so placed as to enjoy the sun till eleven o'clock in the morning.

As the shrubby kinds grow rather fast, they will sometimes fill the pot with their roots, and push them through the opening at the bottom; they must therefore be moved every two or three weeks in the summer, and the fresh roots which are seen pushing through must be cut off. They should also be newly potted twice in the course of the summer: once about a month after they are placed abroad, and again towards the end of August. When this is done, all the roots on the outside of the ball of earth should be carefully pared off, and as much of the old earth removed as can be done without injuring the plants. If they then require a larger pot, they should be planted in one about two inches wider than that from which they have been removed. Some fresh earth should first be placed at the bottom, and on that the plant should be placed in such a manner, that the ball of earth adhering to it may be about an inch below the rim of the pot: it should then be filled up, and the pot a little shaken to settle the earth about the roots: the earth must then be gently pressed down at the top, leaving a little space for water to be given without running over the rim: finally, the plant should be liberally watered, and the stem fastened to a stake, to prevent the winds from displacing the roots before they are newly fixed.

As the branches advance in growth, and new leaves are formed at the tops of them, the lower ones constantly decay: these should be plucked off every week or fortnight; as they are not only unsightly, but injurious to the air about the plants.

The tube-rooted kinds may be increased by parting the roots, which should be done in August: every tuber that has an eye to it will grow. Such as are raised from slips should be planted in May, June, or July, taking only the last year's shoots, from which the lower leaves must be stripped. When planted, give them water, and place them in the shade. In four or five weeks they will have taken root, when they may be so placed as to enjoy the sun till eleven in the morning, and there remain until removed to their winter quarters. The slips chosen for cutting should not be such as bear flowers; and they should be inserted about half their length in the earth.

Many of the Geraniums are annual; and as they are so numerous, it would be well, where there is room but for a few, to select such as are perennial. The cuttings of different species of the Pelargoniums do not all strike root with equal readiness. The following may be readily increased in this manner:

The Multifid-leaved	or	Pelargonium Radula.
The Clammy		Glutinosum.
The Heart-leaved		Cordatum.
The Prickly-stalked		Echinatum.
The Square-stalked		Tetragonum.
The Birch-leaved		Betulinum.
The Ternate		Ternatum.

The shrubby kinds are the most tender; the others require shelter from frost only, and should have free air admitted to them whenever the weather is not very severe: in mild weather, the shrubby kinds also may be permitted to enjoy the fresh air.

In sultry weather the Geraniums should all be watered liberally every evening, with the exception of some few of the Pelargoniums, which are of a succulent nature. Those must be watered sparingly. The succulent ones may be discerned by merely plucking a leaf from them. The season for flowering is generally from April to August.

Those who are curious in Geraniums may see them figured in most of their known varieties, in a very beautiful work, published in numbers, entitled Andrews's Monograph on the Genus Geranium. This work represents them in their full beauty; and, being very finely-coloured, gives you as good an idea of them, as if you had seen the plants themselves. The Elegant, the Magnificent, and the Handsome kinds fully justify their titles. The Geranium

Tricolor Arboreum, or Three-coloured Tree Geranium, is similar, both in the form of the leaves and the flowers, to the Hearts-ease: the flowers are white and red, and uncommonly beautiful. In appearance, it is neither more nor less than a large red and white Hearts-ease. The Oval-leaved Three-coloured Geranium bears a flower somewhat smaller, but of the same form and colour. The Birch-leaved, in all its varieties, is remarkably handsome, with brilliant red flowers. The Wrinkly-leaved has very large and beautiful blossoms: the Sea-green-leaved is an exceedingly elegant and delicate plant: the Heart-leaved particularly luxuriant.

Mr. Andrews observes, that the varieties of the Geranium Citriodorum, or Citron-scented Geranium, are the only ones which make a powerful appeal to the olfactory nerves, without rubbing the leaves. Most of them emit an agreeable odour when lightly rubbed with the finger; and a person approaching a Geranium, almost mechanically rubs or plucks a leaf for its perfume; or, with some species, for its soft velvety surface:

> " And genteel Geranium
> With a leaf for all that come,"

seldom fails of obtaining notice and admiration, however it may be surrounded by the most curious or brilliant exotics.

The Thick-stemmed Geranium is a very singular plant. " This species," says Mr. Andrews, " was found (by Mr. Antoni Pantaleo Hove, in 1785, while Botanical Collector to his Majesty) near five feet high, in the bay of Angra Peguena, on the south-western coast of Africa, in the chasms of a white marble rock, apparently without any earth; for, on pulling up the plant, the roots were several yards in length, naked, and as hard as wire; and appeared to have received nourishment solely from the moisture lodged there during the rainy season, assisted by a little sand drifted by the wind into the cavities. The heat was so intense on these rocks as to blister the soles of the feet; and yet all the Geraniums there were in perfection, being just then their flowering season, about the middle of April*."

The Lance-leaved and Ivy-leaved species are extremely elegant. As there are many kinds of Geranium in estimation, and they differ in being more or less hardy, it may be well to subjoin a little table of those most commonly cultivated here, with these distinctions:

The Geranium Divaricatum	or	Divaricated Geranium.
Carolinianum		Carolina.

These two are annual, should be raised in a hot-bed, kept in the open air during the summer, and will then decay.

The Maculatum	or	Spotted-leaved Geranium.
Sanguineum		Blood-coloured.
Pratense		Meadow.
Phæum		Black red-flowered.
Striatum		Streaked-flowered.
Lancastriense		Lancashire-striped.
Macrorhizum		Long-odorous-rooted.
Palustre		Marsh.
Sylvaticum		Sylvan.
Argenteum		Silvery-leaved.
Nodosum		Knotty.

These are hardy perennial kinds, which, unless in very severe winters, will bear the open air: they should be gently watered every evening in the summer; and three times a week, when not frosty, in the winter. Of the Pelargoniums, demanding winter shelter as directed, are the following:

The Pelargonium Cortusi-folium	or	Cortusa-leaved Pelargonium.
Australe		Botany-bay.

* Andrews' Monograph, on the Genus Geranium, No. 21.

The Pelargonium	Barringtonium or	Barrington Pelargonium.
	Beaufortianum	Beaufort's.
	Betulinum	Birch-leaved.
	Bicolor	Two-coloured.
	Blattarium	Hoary-leaved.
	Citronium	Citron-scented.
	Cordifolium	Heart-leaved.
	Crenatum	Cape Scarlet.
	Formosum	Handsome.
	Fragrans	Fragrant.
	Grandiflorum	Great-flowering.
	Speciosum	Beautiful.
	Tricolor	Three-coloured.
	Tomentosum	Downy.
	Elegans	Elegant.
	Hybridum	Bastard.
	Heterogamum	Lady Coventry's.
	Zonale	Horse-shoe.
	Peltatum	Peltated.
	Inquinans	Scarlet.
	Cucullatum	Hooded.
	Penicillatum	Pencilled.
	Glutinosum	Clammy.
	Angulosum	Marsh-Mallow-leaved.
	Nummulifolium	Coin-leaved.
	Papilionaceum	Butterfly.
	Echinatum	Prickly-stalked.
	Radula	Multifid-leaved.
	Asperum	Rough-leaved.
	Ternatum	Ternate.
	Graveolens	Rose-scented.
	Vitifolium	Vine-leaved, Balm-scented.
	Capitatum	Rose-scented.
	Balsameum	Balsamic.
	Incisum	Gashed.
	Tetragonum	Square-stalked.
	Gibbosum	Gouty.
	Acetosum	Sorrel.
	Denticulatum	Toothed.
	Quercifolium	Oak-leaved.
	Fulgidum	Celandine-leaved.
	Reniforme	Kidney-leaved.
	Fragile	Brittle.

These are chiefly natives of the Cape of Good Hope.

The Geranium which first became familiar to us, that with plain red flowers, is still, and deservedly, a favourite. It may be brought to grow very large by care and attention, and bears an abundance of blossoms. Cowper speaks of it, in describing the inhabitants of the greenhouse:

> "———— Geranium boasts
> Her crimson honours."

Mrs. Charlotte Smith, in her Conversations on Natural History, introduces some lines to a Geranium, which had been carefully nursed.

GERMANDER.

TEUCRIUM.

LABIATÆ. DIDYNAMIA GYMNOSPERMIA.

From Teucer, son of Scamander, and father-in-law of Dardanus, king of Troy.

Of the numerous species of Teucrium, it will suffice to select a few of the most desirable; as,

1. The Many-flowered; red flowers; native of Spain.	
2. The Canadian; yellow; blowing in	August and September.
3. The Virginian; red;	July and August.
4. The Betony-leaved; Persian;	August to October.
5. The Water-leaved; pink, or pale purple; Europe;	July and August.
6. The Common, or Wall; red-purple; Europe, Palestine, and Islands of the Archipelago.	
7. The Shining; yellow; Mount Atlas, South of Europe;	June to September.
8. The Pyrenean; purple and white.	
9. The Poley; yellow or white; South of Europe;	June to July.

The second, third, and fifth kinds will, in mild winters, thrive abroad, even in pots; the fifth must have a very moist soil, and will require more water than the others. The other six kinds must be sheltered from the frosts of winter, and be kept moderately moist.

Most of them may be raised from cuttings planted early in April, and shaded till they have taken root.

Mr. H. Smith, enumerating a variety of flowers which have their origin in the metamorphosis of lovers, &c., adds,

> ——————" *that* baptized
> With Phrygian Teucer's name."

GLOBE-FLOWER.

TROLLIUS.

RANUNCULACEÆ. POLYANDRIA POLYGYNIA.

The botanical name is supposed to be of German origin, and to signify a Magic-flower. It is also called Globe-Ranunculus, and Globe Crow-foot, from the coloured lobes of the calyx being always inflected at the tip, and never expanded, so that they constantly form a complete globe.—*French,* le renoncule de montagne.

THE European Globe-flower is a native of most parts of Europe, growing in moist shady places. It is very common in the north of England; in the south it is found only in gardens. In the northern counties, it is called Locker-goulans, which Mr. Martyn supposes to be a corruption of the Lucker-gowan (Cabbage-daisy) of the Scots. Allan Ramsay makes his young laird seek a chaplet of it for his Katy's brow.

"This splendid flower," says Linnæus, "adorns the pavement of the rustics on festival days." It is a bright yellow flower, blowing in May and June. "In Westmoreland these flowers are collected with great festivity, by the youth of both sexes, at the beginning of June; about which time it is usual to see them return from the woods in an evening, laden with them, to adorn their doors and cottages with wreaths and garlands*.

The Globe-flower may be increased by parting the roots in September, when the leaves begin to decay; but they must not be parted very small, nor oftener than every third year.

The Siberian Globe-flower has paler flowers, and more open, than those of the European species: this also blows in May and June, and may be increased in the same manner; but it requires a soil yet more moist than that does; and the best way to keep it flourishing is to cover the earth with moss, and to water it frequently. They both love the shade; exposure to the sun, and want of water, will soon destroy them. They will bear the open air at all seasons.

GLOBULARIA.

GLOBULARIEÆ. TETRANDRIA MONOGYNIA.

So named by Tournefort, from the flowers growing many together in the form of a little globe, or ball.

THE Montpelier Globularia—in French, *l'arbrisseau terrible; Globulaire turbith;* but in Languedoc, *lou pichot fené*—is a leafy little shrub, the leaves resembling those of the myrtle: the flowers blue. From its medicinal properties it has been named Herb terrible. It flowers from August to November. This plant may be increased by

* Martyn's edition of Miller's Gardener s Dictionary.

cuttings, which should be taken in April, just before it begins to make new shoots. They are usually put into a hot-bed until they have taken root; but, if kept in the house, they will strike very well without. When rooted, they should be inured by degrees to the open air; but must be housed again towards winter, admitting fresh air when not frosty.

The Common Globularia, Globe-daisy, or Blue-daisy, is a native of most other parts of Europe, but not of England: it is called in France, *la boulette.* In Gerarde's, and even in Parkinson's time, it was rarely seen in our gardens. The flower is a beautiful blue, and appears in May and June. This species, the Prickly-leaved, the Wedge-leaved, and the Naked-stalked, may be increased by parting the roots in the September of every second year. They like a moist loamy earth, and will bear the open air.

The Long-leaved Globularia, and the Oriental, may be increased in the same manner, but in other respects must be treated like the first. They all agree in liking the shade, and a frequent supply of water, but they must not have much at a time. All the kinds have blue flowers.

GOAT'S-RUE.

GALEGA.

LEGUMINOSÆ. DIADELPHIA DECANDRIA.

French, rue de chevre; lavanese.—*Italian,* capraggine; ruta capraria; lavanna; lavanese; lavamani; sarracena: in Piedmont, bavarosce.

The Common Goat's-rue grows naturally in Africa, and in many parts of Europe. It has usually blue flowers;

but there are varieties with white and with variegated blossoms. It flowers in June. Some give this species the name of Italian Vetch. It will live in the open air, and must be kept moderately moist.

The Virginian species has red flowers, is less hardy, and must be housed in the winter, admitting fresh air in mild weather. Most of the species are natives of the East or West Indies, and are therefore hot-house plants. Some of these are very handsome. Among them is one called Galega Tinctoria, from which the inhabitants of Ceylon prepare their indigo; and another called Galega Toxicaria, of which the leaves and branches, pounded and thrown into a river or pond, affect the water in such a manner as to intoxicate the fish, and make them float on the surface as if dead. Most of the larger ones recover after a short time, but the greater part of the small fry perish. On account of its intoxicating qualities it is much cultivated in America.

GOLDEN-LOCKS.

CHRYSOCOMA.

CORYMBIFERÆ. SYNGENESIA POLYGAMIA ÆQUALIS.

The botanical name is derived from two Greek words of the above signification. By country-people the English name is corrupted into Goldy-locks.—*French,* crisocome.—*Italian,* crisocoma.

Most of these plants are natives of the Cape of Good Hope, and must therefore be housed during the winter season. They should, however, enjoy the fresh air in mild weather, since they require protection from frost only. They may be increased by cuttings, which should be planted in the summer months, shaded from the sun, and kept moist, and they will easily take root. It will forward them to cover them with a hand-glass while rooting.

The German Golden-locks—called in French *crisocome de Dioscoride, crisocome liniere*—is usually propagated by parting the roots, which should be done in autumn, soon after the roots decay, in order that they may have time to establish themselves in their new stations before the winter. This species will live in the open air. It must be sparingly watered. When touched, it gives out a fine aromatic scent.

The shrubby kinds are in blossom nearly all the year round. The flowers are yellow in all the species.

GOURD.

CUCURBITA.

CUCURBITACEÆ. MONOECIA SYNGENESIA.

French, la courge.—*Italian,* la zucca: at Rome, cucuzza: in the Brescia, suca, co, melona.

The Orange Gourd, and other small varieties, which can be trained round a stake and kept within bounds, will have a pretty effect in a balcony in the summer. They may be sown towards the close of April, and should be watered every evening in dry summer weather. They are annual plants.

The larger species of Gourd are very useful to their countrymen. The Bottle Gourd—named by the French *la calebasse; la gourde; le flacon; la calebasse d'herbe;* and by the Italians *la zucca longa; la mazza d'Ercole* [Hercules's club]—which the Arabians call charrah, is, by the poor, boiled in vinegar and eaten. Sometimes they make it into a kind of pudding, by filling the shell with rice and meat. In Jamaica the shells are in general use as water cups, and frequently serve the negroes and poorer sort of white people for bottles. The largest variety of this species is cultivated for the sake of the shells, which will sometimes contain five, six, or seven gallons. The Warted Gourd—called by the French *le potiron a verrues; la barbarine*—is gathered when half-grown by the Americans, and boiled as a sauce to their meat. The Water Melon—in French, *la pasteque; le melon d'eau; citronelle; concombre citrin:* in Italian, *cocomero; mellone:* in Venice, *anguria:* in the Brescian, *sorgnel*—serves the Egyptians for meat, drink, and medicine, from the beginning of May to the end of July. They are eaten abundantly. When they are very ripe, their juice, mixed with a little rose-water and sugar, forms the only medicine which the common people take in the most ardent fevers.

The Pompion, or Pumpkin,—called in France *le potiron; le pepon; la citronille:* in Italy, *zucca bernoccoluta; popone; poponoino*—which in Europe is considered hard of digestion, is reckoned in the Eastern countries as the most wholesome of all the Gourds. In North America, China, &c. the Squash Gourd—in French, *le pastisson; le bonnet d'electeur*—also is considered as an article of food; and, as it will keep fresh and sweet for several months, is very useful in long voyages. The fruit of the Gourds, when unripe, is generally of a green colour, and, if such a phrase may be allowed, a *very green* green.

> "Then gan the shepherd gather into one
> His straggling goats, and drave them to a foord,
> Whose cærule stream, rombling in pibble-stone,
> Crept under moss, as green as any goord."
>
> Spenser's Virgil's Gnat.

> "Sometimes a poet from that bridge might see
> A nymph reach downwards, holding by a bough
> With tresses o'er her brow;
> And with her white back stoop
> The pushing stream to scoop
> In a green gourd cup, shining sunnily."
>
> HUNT'S NYMPHS.

Cowper appears in the following passage to have confounded the Gourd with the cucumber:

> "To raise the prickly and green-coated gourd,
> So grateful to the palate, and, when rare,
> So coveted; else base and disesteemed,
> Food for the vulgar merely *; is an art
> That toiling ages have but just matured."

GREEK VALERIAN.

POLEMONIUM CÆRULEUM.

POLEMONIACEÆ. PENTANDRIA MONOGYNIA.

Jacob's Ladder; Ladder to Heaven.

THIS plant has no affinity to the valerian: it has only some little resemblance in the shape of the leaves. The flowers are pretty, blue or white, and open about the end of May. It is a native of Asia and the North of Europe. The seeds may be sown in spring, in a fresh light soil, not very rich. At Michaelmas they may be transplanted into separate pots, of a middle size: or they may be in-

* A new species of Gourd has been very lately introduced from Persia under the name of vegetable marrow; the flesh, when not fully ripe, having a peculiar softness, and, when peeled and boiled, resembling the buttery quality of the beurré pears. It is easily cultivated, and promises to be a great acquisition to our tables.

creased by parting the roots in autumn. The earth should be moderately moist, but never *wet*.

GUELDER-ROSE.

VIBURNUM OPULUS.

CAPRIFOLIEÆ. PENTANDRIA TRIGYNIA.

Elder-rose; Rose-elder; Snowball-tree.

THIS elegant shrub is a variety of a species of viburnum called Water-elder, and delights in a moist soil. The name of Snowball-tree is so appropriate as naturally to suggest itself to the mind; and I have more than once heard it remarked by persons who knew it only by its more general title of Guelder-rose, that it *should* have been called the Snowball-tree.

It has, at first sight, the appearance of a little maple-tree that has been pelted with snowballs; and we almost fear to see them melt away in the sunshine. This beautiful snowball of summer continues, however, to adorn the green leaves, which so finely contrast with its whiteness, for two or three successive months, first appearing towards the end of May.

When kept in pots, the Guelder-rose will require watering every evening in dry summer weather. Being a native of North America, it will bear our climate very well; but it will be important, when in blossom, to shelter it from heavy rains, which would be apt partially to *thaw* these delicate flowers.

Cowper, who loved his garden, and found new pleasure

in transplanting his flowers into his poems, describes the Guelder-rose as

———————————" tall,
And throwing up into the darkest gloom
Of neighbouring cypress, or more sable yew,
Her silver globes, light as the foamy surf
That the wind severs from the broken wave."

HAWTHORN.

CRATÆGUS OXYACANTHA.

ROSACEÆ. ICOSANDRIA DIGYNIA.

French, l'aubepine; l'épine-blanche; la noble épine; le senellier.—*Italian,* bianco-spino; amperlo; bagaia.—*English,* Hawthorn, from the Anglo-Saxon, hægthorn; Whitethorn; Quick; May-bush.

FEW trees exceed the Common Hawthorn in beauty, during the season of its bloom. Its blossoms have been justly compared to those of the myrtle: they are admirable also for their abundance, and for their exquisite fragrance. This shrub usually flowers in May; and being the handsomest then, or perhaps at any time, wild in our fields, has obtained the name of May, or May-bush. The country-people deck their houses and churches with the blossoms on May-day, as they do with holly at Christmas.

" Youth's folk now flocken everywhere,
To gather May-buskets and smelling breere;
And home they hasten the posts to dight,
And all the kirk-pillars ere day-light,
With hawthorne buds, and sweet eglantine,
And girlonds of roses, and sops-in-wine."

SPENSER'S SHEPHERD'S CALENDAR.

There are many species of Hawthorn. India has its Hawthorn: America, China, Siberia, have each their Hawthorn: several are Europeans: but our own British shrub yields to none of them. It is very common in every part of England; is to be seen in every hedge:

> "And every shepherd tells his tale
> Under the hawthorn in the dale."
>
> MILTON, L'ALLEGRO.

We must not, however, let our fancies run so riot, as to suppose that the poet here intends that we should conceive a beautiful and youthful nymph sitting by the shepherd's side, to whom he is pouring forth his fond tale of love; for, in very truth, the real image present in the poet's mind was simply that of a shepherd telling his tale, or, in unpoetic language, counting his sheep, as he lies extended in the shade of this tree; and to those who take pleasure in a country life, and rural associations, perhaps this image will appear scarcely less poetical, or less pleasing, than the former interpretation, which many readers give to this passage at first sight.

This tree not only delights our senses with its beauty and perfume, and affords a cooling shade in sunny fields, a benevolence for which it has been celebrated by many of our best poets, but it also harbours the little birds which cheer us with their joyous music. The thrush, and many others, feed in winter on its berries, the bright scarlet haws. A decoction of the bark yields a yellow dye: the wood is used for axle-trees and tool-handles. "The root of an old Thorn," says Evelyn, "is excellent for boxes and combs. When planted single, it rises with a stem big enough for the use of the turner; and the wood is scarcely inferior to box."

The Glastonbury variety, commonly called the Glastonbury Thorn, usually flowers in January or February; but it is sometimes in blossom on Christmas-day. In many countries the peasants eat the berries of the Hawthorn; and the Kamschatkadales make a wine from them.

The Hawthorn will grow many years in a pot or tub, and require no other care than watering it occasionally in dry weather, and removing it into a larger pot as it outgrows the old one.

The scent of the May-blossom is proverbially sweet. How much is said in praise both of its beauty and sweetness in the following couplet!

"A bush of May-flowers with the bees about them;
Ah, sure no tasteful nook would be without them!"
KEATS.

Chaucer frequently speaks of the Hawthorn:

"There sawe I growing eke the freshe hauthorne
In white motley, that so sote doeth ysmell."
COMPLAINT OF THE BLACK KNIGHT.

In the celebration of May-day, in the Court of Love, he says:

"And furth goth all the Courte both most and lest
To fetche the flouris freshe, and braunch and blome,
And namely hauthorne brought both page and grome,
With fresh garlandis, party blew and white,
And than rejoysin in their grete delight."

"Amongst the many buds proclaiming May,
(Decking the fields in holiday's array,
Striving who shall surpasse in bravery)
Marke the faire blooming of the hawthorne-tree;
Who, finely cloathed in a robe of white,
Feeds full the wanton eye with May's delight;
Yet for the bravery that she is in
Doth neyther handle carde nor wheele to spin,
Nor changeth robes but twice, is never seene
In other colors than in white or greene.
Learn then content, young shepherd, from this tree,
Whose greatest wealth is Nature's livery."

"All the trees are quaintly tyred
With greene buds of all desired;
And the hauthorne every day
Spreads some little show of May.

See the primrose sweetly set
By the much-loved violet,
All the bankes doe sweetly cover
As they would invite a lover
With his lass, to see their dressing,
And to grace them by their pressing."

W. Browne

" 'Tis May, the Grace,—confess'd she stands
By branch of hawthorn in her hands:
Lo! near her trip the lightsome dews,
Their wings all tinged in iris hues;
With whom the powers of Flora play,
And paint with pansies all the way."

Warton.

Philips, in his Letter from Copenhagen, beautifully describes the appearance of the Hawthorn in the winter:

" In pearls and rubies rich the hawthorns show,
While through the ice the crimson berries glow."

There is a beautiful address to the Hawthorn in the poems of Ronsard. The following version*, which is from the pen of the Rev. Mr. Cary, is so faithful, and so happy, that the French poet will suffer no injustice if we quote the translation only:

" Fair hawthorn flowering,
With green shade bowering
Along this lovely shore;
To thy foot around
With his long arms wound
A wild vine has mantled thee o'er.

" In armies twain,
Red ants have ta'en
Their fortress beneath thy stock:
And in clefts of thy trunk
Tiny bees have sunk
A cell where honey they lock.

* See " Notices of the Early French Poets," in the London Magazine, vol. v. p. 511.

"In merry spring-tide,
  When to woo his bride
The nightingale comes again,
  Thy boughs among
  He warbles his song,
That lightens a lover's pain.

"'Mid thy topmost leaves
  His nest he weaves
Of moss and the satin fine,
  Where his callow brood
  Shall chirp at their food,
Secure from each hand but mine.

"Gentle hawthorn, thrive,
  And, for ever alive,
Mayst thou blossom as now in thy prime;
  By the wind unbroke,
  And the thunderstroke,
Unspoiled by the axe of time."

The following lines by another French poet, Olivier de Magny, addressed to Ronsard's servant, present a most delightful picture:

"And if he with his troops repair
  Sometimes into the fields,
Seek thou the village nigh, and there
  Choose the best wine it yields.
Then by a fountain's grassy side,
  O'er which some hawthorn bends,
Be the full flask by thee supplied,
  To cheer him and his friends."

London Magazine, vol. v. p. 159.

HEART'S-EASE.

VIOLA TRICOLOR.

VIOLÆ. SYNGENESIA MONOGYNIA.

French, herbe de la Trinité; pensees [thoughts].—*Italian*, flammola [little flame]; viola farfalla [butterfly violet]; viola segolina [winged violet]; fior della Trinita; suocera e nuora [mother-in-law and daughter-in-law]. The Greeks have named it phlox [a flame].

This beautiful flower is a native of Siberia, Japan, and

many parts of Europe. It is a general favourite, as might be supposed from the infinity of provincial names which have been bestowed upon it from its beautiful colours:—

Love in Idleness.	Jump up and kiss me.
Live in Idleness.	Look up and kiss me.
Call me to you.	Kiss me ere I rise.
Cull me to you.	Kiss me behind the Garden-gate.
Three Faces under a Hood.	Pink of my John.
Herb Trinity.	Flower of Jove.

And Flamy, because its colours are seen in the flame of wood.

It is a species of violet, and is frequently called the Pansy-violet, or Pansy, a corruption of the French name, *pensées.*

The smaller varieties are scentless, but the larger ones have an agreeable odour. Drayton celebrates its perfume by the flowers with which he compares it in this respect; but then, to be sure, his is an Elysian Heart's-ease:

"The pansy and the violet, here,
As seeming to descend
Both from one root, a very pair,
For sweetness do contend.

"And pointing to a pink to tell
Which bears it, it is loth
To judge it; but replies, for smell
That it excels them both.

"Wherewith displeased they hang their heads,
So angry soon they grow,
And from their odoriferous beds
Their sweets at it they throw."

The Heart's-ease has been lauded by many of our poets; it has been immortalised even by Shakspeare himself; but no one has been so warm and constant in its praise as Mr. Hunt, who has mentioned it in many of his works. In the Feast of the Poets, he entwines it with the Vine and the Bay, for the wreath bestowed by Apollo upon Mr. T. Moore. In the notes to that little volume, he again speaks of this flower, and I do not know that I can do better than steal a few of its pages to adorn this.

" It is pleasant to light upon an universal favourite, whose merits answer one's expectation. We know little or nothing of the common flowers among the ancients; but as violets in general have their due mention among the poets that have come down to us, it is to be concluded that the Heart's-ease could not miss its particular admiration, —if indeed it existed among them in its perfection. The modern Latin name for it is *flos Jovis,* or Jove's flower,—an appellation rather too worshipful for its little sparkling delicacy, and more suitable to the greatness of an hydrangea or to the diadems of a rhododendron.

> " Quæque per irriguas quærenda Sisymbria valles
> Crescunt, nectendis cum myrto nata coronis;
> Flosque Jovis varius, folii tricoloris, et ipsi
> Par violæ, nulloque tamen spectatus odore.
>
> RAPINI HORTORUM, lib. i.

> " With all the beauties in the vallies bred,
> Wild mint, that 's born with myrtle crowns to wed,
> And Jove's own flower, that shares the violet's pride,
> Its want of scent with triple charm supplied.

" The name given it by the Italians is *flammola,* the little flame;—at least, this is an appellation with which I have met, and it is quite in the taste of that ardent people. The French are perfectly *aimâble* with theirs:—they call it *pensée,* a thought, from which comes our word Pansy:—

" 'There 's rosemary,' says poor Ophelia; 'that 's for remembrance; —pray you, love, remember;—and there is pansies,—that 's for thoughts.' Drayton, in his world of luxuries, the Muse's Elysium, where he fairly stifles you with sweets, has given, under this name of it, a very brilliant image of its effect in a wreath of flowers;—the nymph says,

> " Here damask roses, white and red,
> Out of my lap first take I,
> Which still shall run along the thread;
> My chiefest flow'r this make I.
> Amongst these roses in a row,
> Next place I pinks in plenty,
> These double-daisies then for show;
> And will not this be dainty?

The pretty pansy then I 'll tye,
Like stones some chain enchasing;
The next to them, their near ally,
The purple violet placing.
NYMPHAL 5th.

" Milton, in his fine way, gives us a picture in a word,

" ——————the pansy *freak'd* with jet.

" Another of its names is Love-in-idleness, under which it has been again celebrated by Shakspeare, to whom we must always return, for any thing and for every thing;—his fairies make potent use of it in the Midsummer-Nights' Dream. The whole passage is full of such exquisite fancies, mixed with such noble expressions and fine suggestions of sentiment, that I will indulge myself, and lay it before the reader at once, that he may not interrupt himself in his chair:—

OBERON. My gentle Puck, come hither:—thou rememberest,
Since once I sat upon a promontory,
And heard a mermaid, on a dolphin's back,
Uttering such dulcet and harmonious breath,
That the rude sea grew civil at her song,
And certain stars shot madly from their spheres
To hear the sea-maid's music?
PUCK. I remember.
OBERON. That very time I saw (but thou couldst not,)
Flying betwixt the cold earth and the moon,
Cupid all arm'd:—a certain aim he took
At a fair vestal, throned by the west,
And loosed his love-shaft smartly from his bow,
As it should pierce a hundred thousand hearts:
But I might see young Cupid's fiery shaft
Quench'd in the chaste beams of the watery moon;
And the imperial votaress pass'd on,
In maiden meditation, fancy free.
Yet mark'd I where the bolt of Cupid fell:—
It fell upon a little western flower,—
Before, milk-white,—now purple with love's wound,—
And maidens call it Love-in-idleness.
Fetch me that flower,—the herb I show'd thee once:
The juice of it, on sleeping eyelids laid,
Will make or man or woman madly dote
Upon the next live creature that it sees.

Fetch me that herb; and be thou here again
Ere the leviathan can swim a league.
 PUCK. I'll put a girdle round about the earth
In forty minutes.
Act II. Sc. 2.

"Besides these names of Love-in-idleness, Pansy, Heart's-ease, and Jump-up-and-kiss-me, the tri-coloured violet is called also, in various country places, the herb Trinity, Three-faces-under-a-hood, Kiss-me-behind-the-garden-gate, and Cuddle-me-to-you, which seems to have been altered by some nice apprehension into the less vivacious request of Cull-me-to-you.

"In short, the Persians themselves have not a greater number of fond appellations for the rose, than the people of Europe for the Heart's-ease. For my part, to whom gaiety and companionship are more than ordinarily welcome on many accounts, I cannot but speak with gratitude of this little flower,—one of many with which fair and dear friends have adorned my prison-house, and the one which outlasted all the rest."

Mr. Hunt again mentions this flower with great praise in his Descent of Liberty; where, after sketching in vivid colours a number of beautiful flowers, he thus finishes the floral picture:

"And as proud as all of them
Bound in one, the garden's gem,
Heart's-ease, like a gallant bold,
In his cloth of purple and gold."

In his enumeration of the flowers in blossom, in his History of the Months, too fond of the Heart's-ease even to name it without a passing commendation, he calls it the Sparkler; a name which it so truly deserves, that it might well be added to those it now bears; in which it already surpasses a Spanish grandee.

Spenser includes the Heart's-ease among the flowers to be strown before Queen Elizabeth:

"Bring hither the pink and purple columbine,
 With gilliflowers:
Bring coronations and sops-in-wine,
 Worn of paramours.

Strow me the ground with daffadowndillies,
And cowslips, and king-cups, and loved lilies:
The pretty pawnce,
And the chevisaunce,
Shall match with the fair floure-de-lice."

Pansies make a part of the wreaths brought by the grateful shepherds to the nymph Sabrina,

"That with moist curb sways the smooth Severn stream.

* * * * *

* * * * *

——— the shepherds at their festivals
Carol her goodness, loud in rustic lays,
And throw sweet garland wreaths into her stream
Of pansies, pinks, and gaudy daffodils."

"The delicacy of its texture, and the vivacity of its purple, are inimitable," says the Countess, in *La Spectacle de la Nature*. "The softest velvets, if set in competition with this flower, would appear to the eye as coarse as canvas."

Yet, in another part of this work, the same flower is represented as an humble one which makes no figure, but diffuses an agreeable odour.

It has already been observed, that only the larger kinds have any scent: thus many persons, judging from the smaller, have thought them all scentless. The difference of opinion on this point may be seen in several of the above quotations.

Dryden, in his translation of a passage in Virgil's Pastorals where the poet speaks of sweet herbs in general, introduces the Pansy; but expressly to distinguish it from a fragrant plant:

"Pansies to please the sight, and cassia sweet to smell."

There is a species of Heart's-ease called the Great Flowering—a native of Switzerland, Dauphiny, Silesia, and the Pyrenees—which is very similar to the common kind, but that it has more yellow in it; and another, called the Yellow Mountain Heart's-ease, of British growth, which, notwithstanding the name it bears, is as often purple and yellow, or even purple alone, as all yellow.

It would be an impertinence to attempt to describe the Heart's-ease; therefore let us proceed at once to the treatment of this little favourite. The roots may be purchased so cheaply, and the flowers of these will be so much finer than any that are sown at home, that this will be much the best way of procuring them. At a nursery, or at Covent-Garden flower-market, six or more may be had for a shilling, all of them covered with flowers and buds. They love the sun, but must be liberally watered every evening to replenish the moisture, which it will consume.

It is said somewhere that the Heart's-ease is sacred to Saint Valentine. It must be confessed to be a choice worthy of that amiable and very popular saint; for the flower, like love, is painted in the most brilliant colours, is full of sweet names, and grows alike in the humblest as well as the richest soils. Another point of resemblance, too, may be added, that where once it has taken root, it so pertinaciously perpetuates itself, that it is almost impossible to eradicate it.

HEATH.

ERICA.

ERICINEÆ. OCTANDRIA MONOGYNIA.

In some parts of England Heath is called Ling, probably from the Danish, lyng: in Shropshire, Grig, from the Welsh, grûg: in Scotland, Hather, or Heather; which, like the English Heath, is derived from the Anglo-Saxon, hæth.—*French,* la bruyere; lande; la brande; le petrole.—*Italian,* Erica, Macchia.

"This genus," says Mr. Martyn, "has, within the compass of a few years, risen from neglect to splendour. Every one remembers that Mr. Pope marks it with contempt, at the same time that he celebrates the colour of the flowers:

'E'en the wild heath displays its purple dyes.'

" Mr. Millar, so late as the year 1768, makes mention of no more than five; four of which, as being wild, he consigns to oblivion."

There are now some hundred species; of which many require the heat of a stove, and very few of them are hardy enough to bear this climate unsheltered. The species from the Cape are many of them very beautiful.

All Europe, and the temperate parts of the vast Russian empire, abound with Heath. The Common Heath, which is little regarded in warmer climates, is used for a variety of purposes in the bleak and barren Highlands of Scotland, and in other northern countries. The poor people use it as thatch for the roofs of their huts, and construct the walls with alternate layers of heath, and a kind of cement made of black earth and straw. The hardy Highlanders frequently make their beds with it. In the Western Isles it affords a dye. Woollen cloth boiled in alum water, and afterwards in a strong decoction made from the green tops and flowers of this plant, becomes of a beautiful orange-colour. Brettius relates, that a kind of ale brewed from these young tops was much used by the Picts: and it is said to be still an ingredient in the beer in some of the Western Isles. In many parts of Great Britain besoms are made of this Heath; and it is an excellent fuel. The flowers are either a kind of rose-colour slightly tinged with purple, or they are quite white. Bees collect a great quantity of honey from them.

This kind, the Fine-leaved, the Cornish, the Ciliate-leaved, the Many-flowered, the Irish, and the Cross-leaved, are hardy, and will bear the open air. The latter is very handsome, and blows twice in the year.

The White Three-flowered Tree Heath, the Portugal, and the Purple Mediterranean, are not very tender, but must be sheltered in severe frost.

The following kinds may stand in the open air in the summer, and be housed about the end of September:

The Banksii.	The Donnea.	The Ventricosa.
Linnæana.	Monsoniæ.	Patersoni.
Caffra.	Mammosa.	Incarnata.
Cerinthoides.	Vestita.	Cubica.
Coccinea.	Tiaræflora.	Nudiflora.
Comosa.	Humeana.	Mucosa.
Formosa.	Colorans.	Sparrmanni.
Grandiflora.		

These are all beautiful; but an attempt to enumerate all that are so would be vain. The earth about the roots of a Heath should be as little stirred as possible; and they should be seldom and sparingly watered.

————— " The Erica here,
That o'er the Caledonian hills sublime
Spreads its dark mantle, (where the bees delight
To seek their purest honey) flourishes,
Sometimes with bells like amethysts, and then
Paler, and shaded like the maiden's cheek
With gradual blushes—other while, as white
As rime that hangs upon the frozen spray.
Of this, old Scotia's hardy mountaineers
Their rustic couches form; and there enjoy
Sleep, which, beneath his velvet canopy,
Luxurious Idleness implores in vain."

Mrs. C. Smith.

The Highland Heath-bed is pleasantly described in the novel of Rob Roy:—" While the unpleasant ideas arising from this suggestion counteracted the good effects of appetite, welcome, and good cheer, I remarked that Rob Roy's attention had extended itself to providing us better bedding than we had enjoyed the night before. Two of the least fragile of the bedsteads, which stood by the wall of the hut, had been stuffed with heath, then in full

flower, so artificially arranged that the flowers, being uppermost, afforded a mattress at once elastic and fragrant. Cloaks, and such bedding as could be collected, stretched over this vegetable couch, made it both soft and warm."
—Rob Roy, chap. 20.

HELIOTROPE.

HELIOTROPIUM.

HELIOTROPEÆ. PENTANDRIA MONOGYNIA.

The word Heliotrope is derived from two Greek words, signifying the *sun*, and *to turn*: the leaves or flowers of this plant having been supposed to turn with the sun. For the same reason it is called Turnsole, which is, indeed, only a French translation of the Greek name. The Italians call the common European species, orologio dei cortegiani [courtiers' dial]; eliotropio; verrucaria.—*French,* l'heliotrope commun; l'herbe aux verrues; le verrucaire, from its use in taking off warts; le tournesol.

The Peruvian Heliotrope is chiefly admired for its fragrance: it is an elegant and delicate plant, but not showy. The blossom is very small, of a pale blue, often inclining to white, and shedding an almond-like perfume, which has gained the plant general favour. It should be housed in autumn, before the weather becomes sharp. If in a pure atmosphere, it will flower great part of the winter: but, though carefully guarded from cold, it must be placed where the air is refreshed by frequent ventilation. In dry summer weather it should have a little water every evening: in winter, not more than twice or thrice a week, and very little at a time.

The Indian, Glaucous, and Small-flowered kinds are annual plants—natives of the West Indies—flowering in June, July, and August; and may be treated in the same manner as the Peruvian.

The Canary Heliotrope (to which the gardeners, it is not known for what reason, have given the name of Madame de Maintenon) is not so tender*: it must be sheltered from frost, but should have plenty of fresh air in mild weather. Cuttings from this species, planted in summer, placed in the shade, and regularly watered, will take root in five or six weeks in the open air.

The Trailing Heliotrope, from the Cape of Good Hope, and the European, are hardy annual plants, which may be sown in September or October, kept in the open air, watered as the others, and will flower in July and August.

The Heliotrope is said to owe its existence to the death of Clytie, who pined away in hopeless love of the god Apollo:

> "She with distracted passion pines away,
> Detesteth company; all night, all day,
> Disrobed, with her ruffled hair unbound,
> And wet with humour, sits upon the ground:
> For nine long days all sustenance forbears;
> Her hunger cloy'd with dew, her thirst with tears:
> Nor rose; but rivets on the god her eyes,
> And ever turns her face to him that flies.
> At length, to earth her stupid body cleaves:
> Her wan complexion turns to bloodless leaves,
> Yet streak'd with red: her perish'd limbs beget
> A flower, resembling the pale violet;
> Which with the sun, though rooted fast, doth move;
> And being changed, changeth not her love."
>
> SANDYS'S OVID, Fourth Book.

* If the name, "Madame de Maintenon," is of French origin, it was perhaps a piece of flattery to Louis the Fourteenth, as the sun to which his favourite lady always turned her eyes.

HELMET-FLOWER.

SCUTELLARIA.

LABIATÆ. DIDYNAMIA GYMNOSPERMIA.

The common European species is also called the hooded willow herb; skull cap.—*French,* la toque [skull cap]; centaurée bleue [blue centaury]; tertianaire, from its use in curing the tertian ague.—*Italian,* terzanaria; scodelletta [skull cap].

THE kinds of Helmet-flower most generally cultivated in our gardens, are, the Oriental, with yellow flowers, blowing in May, June, and July; the Alpine, which has a violet-coloured flower with a white lip—a native of Cochin-China, and of several parts of Europe; the Florentine, with large violet-coloured flowers; and the Tall Helmet-flower, with purple blossoms, from the Levant. They may be sown in autumn, in separate pots, in a dry, poor earth; must be sparingly watered, and stand in the open air. They will not last many years: the Oriental kind will not bear transplanting.

HELONIAS.

MELANTHACEÆ. HEXANDRIA TRIGYNIA.

THE Helonias is a native of North America. The flowers are handsome; their colour white or red, according to the species: they may be increased by offsets taken from the roots in autumn. They like a light, fresh soil, and are hardy enough to thrive in the open air. The roots must not be removed oftener than every third year. The earth should be moderately moist.

HEPATICA.

ANEMONE HEPATICA.

RANUNCULACEÆ. POLYANDRIA POLYGYNIA.

Called formerly the noble liverwort.—*French,* l'anémone hépatique; l'hépatique des jardins.—*Italian,* anemone fegatella.

THE Hepatica is a Swiss species of the anemone: there are many varieties, both single and double, varying in colour, and generally blowing in great profusion in February and March. The flower lies a year within the bud, complete in all its parts. The double flowers last longer than the single, and are much handsomer. They thrive best when exposed only to the morning sun; cold does not injure them. They should be kept moderately moist, and may be increased by parting the roots, which should be done in March, when they are in flower; but not oftener than every third or fourth year. Frequent removal weakens, and sometimes destroys them.

A remarkable instance is recorded of change of colour in these flowers. Some roots of the Double Blue Hepatica being sent from a garden in Tothill-fields to another at Henley upon Thames, when they came to blossom produced white flowers, owing to the difference of the soil: but it is yet more curious, that being returned to their former station, they resumed their original blue colour.

HIBISCUS.

MALVACEÆ. MONADELPHIA POLYANDRIA.

The China Rose and the Changeable Rose are species of the Hibiscus; and the former is reckoned the most beautiful of this handsome genus. It is called by the Indians the Gem of the Sun. With them it grows to a moderately-sized tree; here it is but a shrub. Its native country has not been correctly ascertained, but it is very common both in China and Cochin-China for garden-hedges, as well as in their gardens, and in those of the East Indies. The Indians make these beautiful flowers into festoons and garlands on all occasions of festivity, and even in their sepulchral rites. They are also put to a very different and humble use: that of blacking shoes, whence it has been named the Shoe-flower. The women blacken their hair and eye-brows with these roses, which blow nearly all the year round. There is a variety with white flowers.

The *Hibiscus Mutabilis*, or Changeable Rose, has leaves as large as those of the vine. The flower first opens white, from which it changes to rose-colour, and finally to purple. In the West Indies, all these changes take place in the same day; but here they occupy the space of a week. This plant is a native of the East Indies; from whence the French, who call it *la fleur d'une heure*, carried it to their settlements in the West Indies. It blows in November.

A third species of Hibiscus is the Venice Mallow, or Hibiscus Trionum, one of the very few species belonging to this beautiful genus which may be raised and preserved without the aid of a stove. It is a native of Italy and Austria, bears a purple and yellow flower, and has long been known in our English gardens by the name of the Venice Mallow, Mallow of an Hour, Bladder Ketmia, Bladder Hibiscus, or Good Night at Noon. "But," says Gerarde, "it should rather be Good Night at Nine; for this beautiful flower opens at eight in the morning, and, having received the beams of the sun, closes again at nine."

"Ovid," continues he, "in speaking of the Adonis flower, is thought to describe the Anemone, or Wind-flower, which we rather deem to be this quick-fading Mallow; for it is evident that Adonis flower, and all under the title of Wind-flower, last more than one day; but this is so frail that it scarcely lasts an hour. Bion of Smyrna, an ancient poet, says in his epitaph on Adonis, that the Wind-flower sprung from Venus's tears while she was weeping for Adonis; but, doubtless, the plant was mistaken by the poet, considering the fragility of the flower and the matter whereof it sprung, that is a woman's tears, which last not long; as this flower, *flos horæ*, or Flower of an Hour."

Notwithstanding the facetiousness of the good Gerarde, however, the Venice Mallow must be contented with her own natural parentage, for that of the Anemone is too well established to allow of her being superseded.

Miller says, "that in fine sunny weather this flower will remain open the whole day; that in wet weather it will not open at all; but, when very fine, has been observed not to close until half-past six in the evening."

The first two plants must have very little water in winter, and not a great deal in summer: they are tender, and will always be better in the house, placed near an open window in the summer, and kept pretty warm in the winter. An inhabited room will answer very well for them.

The Venice Mallow, or Bladder Ketmia, being more hardy, may be sown about the end of March; and, when three or four inches high, may be potted separatelv, and occasionally watered; often, if the weather be very dry.

HOLLYHOCK.

ALTHÆA.

MALVACEÆ. MONADELPHIA POLYANDRIA.

This plant is sometimes called the Garden Mallow—*French,* Rose-d'outre-Mer, or Beyond-sea-Rose; sometimes corrupted into la rose trèniere, or la rose tremiere; la mauve rose; la passe rose.—*Italian,* alcea rosea; rosa Cinese.

THE Double Chinese Hollyhock is a very handsome plant, and continues in beauty during July, August, and September. The seeds may be sown early in April, half an inch deep. When the plants have put out six or eight leaves, they should be transplanted into separate pots, and they will require them pretty large, at least a foot in diameter. Until they are well-rooted, they must be watered daily; afterwards, three times a week will suffice. They should be housed in winter, admitting fresh air in mild weather; and, while in the house, should have only water enough to keep the earth from parching. They will last two or three years. They must be supported with stakes to prevent the wind from breaking the stems. These plants may also be raised from cuttings of the young stalks, taken in summer, about six inches in length; they should be inserted half their depth, and if a glass be placed over them, it will facilitate their rooting: plants so raised will flower early in the following summer. The Hollyhock is used, in some parts of France, as we use Hawthorn and Privet, to divide gardens and vineyards. The flowers are said to furnish a large portion of honey-juice to bees:

> " And from the nectaries of Hollyhocks
> The humble bee, e'en till he faints, will sip."
>
> H. SMITH.

HONESTY.

LUNARIA.

CRUCIFERÆ. TETRADYNAMIA SILICULOSÆ.

French, la lunaire [moon-wort]; satinée, satin-blanc [white satin]; passe-satin [slip satin]; medaille [the medal]; and herbe aux lunettes [spectacle herb].—*Italian,* lunaria.—*German* and *Dutch,* a variety of names, most of which have a signification equivalent to silver-bloom.—*English,* Honesty; Moon-wort; Penny-flower; Money-flower; White Satin-flower; and Silver-plate.

THE Italian name Lunaria, and the English name Moon-wort, were given to this plant from the form of the seed-vessel, which resembles that of the full moon. Honesty has a reference to the transparency of the partition of the seed-vessel; Silver-bloom, Satin-flower, &c. to its smoothness and glossiness. The French name, *Herbe aux Lunettes,* is very appropriate, for this part of the plant is certainly very like the oval glass of a pair of spectacles.

The lower flowering stalks of this plant have few flowers and some leaves; but the upper ones have many flowers and not any leaves. The plant is chiefly remarkable for the partition of the seed-vessel from which it is named; and, when this is perfectly ripe, the branches are frequently dried, and placed with Amaranths, Xeranthemums, &c. over the chimney-piece in the winter-time.

The perennial Honesty grows naturally in many parts of Europe: the biennial kind is a native of Germany. They flower in May and June; the blossoms are purple or white, usually without scent; but there is one variety of which the scent is very agreeable. The annual species is a native of Egypt.

They may be raised from seed, which should be sown in the autumn, singly, and should not be afterwards transplanted. They will grow in any soil, and in the open air, and love the shade.

This is mentioned by Chaucer as one of the plants used in incantation:

> "And herbes coude I tell eke many on,
> As egremaine, valerian, and lunarie,
> And other swiche, if that me list to tarie,
> Our lampes brenning bothe night and day,
> To bring about our craft if that we may,
> Our fourneis eke of calcination,
> And of wateres albification."

Drayton also speaks of its magical virtues:

> "Enchanting Lunary here lies,
> In sorceries excelling."

HONEY-SUCKLE.

LONICERA.

CAPRIFOLIEÆ. PENTANDRIA MONOGYNIA.

This botanical name was given by Plumier, in honour of Adam Lonicer, a physician of Frankfort.—*French,* chevre feuille des bois [wood honeysuckle]; maire sauvage; pantacouste sauvage; both signifying wild honey-suckle.—*Italian,* caprifoglio; madreselva; vincibosco; legabosco; periclimeno.—*English,* Honey-suckle; Suckling; Caprifoly; Woodbine, or Woodbind.

Few flowers have been more admired or cultivated than

the Honey-suckle. The European languages seem to vie with each other in the number of names bestowed on this beautiful favourite; but the German has outstript all the rest, in reference to this plant as well as most others; the greatest part of them having in that language at least a dozen common names. There are many species of Honey-suckle, and of most of the species several varieties; but as they are invariably beautiful, any that can be reared with success in a pot will be valuable. They will live in the open air, and in dry summer weather should be liberally watered every evening.

The common English Honeysuckle is also called Woodbinde, or Woodbine:

> "So doth the woodbine, the sweet honeysuckle
> Gently entwist."

"Shakspeare seems here to have distinguished the Honeysuckle from the Woodbine," says Mr. Martyn. Yet, in Much Ado About Nothing, he uses either name indiscriminately:

> "And bid her steal into the pleached bower
> Where honeysuckles ripened by the sun
> Forbid the sun to enter."
> * * * * *
> ————————"Beatrice, who e'en now
> Is couched in the woodbine coverture."

"Milton," observes Mr. Martyn, "seems to have mistaken it, when he gives it the name of Eglantine, and distinguishes it from Sweet-briar, since the Sweet-briar is itself the Eglantine:

> "Through the sweet-briar, or the vine,
> Or the twisted eglantine."

Shakspeare justly distinguishes the two:

> "I know a bank whereon the wild thyme blows,
> Where oxlip, and the nodding violet grows:

O'ercanopied with luscious woodbine,
With sweet musk-roses, and with eglantine."

In Comus, Milton speaks of it by its proper name.

" I sat me down to watch upon a bank
With ivy canopied, and interwove,
And flaunting honeysuckle."

And by the name of Woodbine in his Paradise Lost:

" Let us divide our labours, thou where choice
Leads thee, or where most needs, whether to wind
The woodbine round this arbour, or direct
The clasping ivy where to climb, while I,
In yonder spring of roses, intermixed
With myrtle, find what to redress till noon."

The rambling nature of the Honeysuckle is usually its chief character in poetry:

" You'll find some books in the arbour: on the shelf
Half hid by wandering honeysuckle."

BARRY CORNWALL'S FALCON.

——————" the poplar there
Shoots up its spire, and shakes its leaves i'the sun
Fantastical, while round its slender base
Rambles the sweet-breathed woodbine ———."

BARRY CORNWALL.

" And there the frail-perfuming woodbine strayed
Winding its slight arms 'round the cypress bough,
And, as in female trust, seemed there to grow,
Like woman's love midst sorrow flourishing."

BARRY CORNWALL.

Cowper evidently alludes here to the wild Woodbine in our hedges, which is sometimes nearly white:

" Copious of flowers, the woodbine pale and wan,
But well compensating her sickly looks
With never cloying odours, early and late."

Chaucer repeatedly introduces the Woodbine, for arbours, garlands, &c.; and in one passage makes it an emblem of fidelity, like the violet:

"And tho' that were chapèlets on their hede
Of fresh wode-bind be such as never were
To love untrue in word, in thought, ne dede,
But ay stedfast, ne for plesaunce ne fere,
Tho' that they shudde their hertis all to tere,
Woud never flit, but evir were stedfast
Till that ther livis there assunder brast."

THE FLOURE AND THE LEAFE.

The Honeysuckle varies in colour, not only the different species, but even different blossoms on the same tree: some are beautifully dashed with white and crimson; others are variegated with shades of purple, or yellow, or both: thus its colour is seldom described. Philips notices its colour in one of his pastorals:

"And honeysuckles of a purple dye."

Varying as it does in colour, all the different kinds are brought at once before us by this half line—from the story of Rimini:

---------"the suckle's streaky light."

HOTTENTOT CHERRY.

CASSINE MAUROCENIA.

RHAMNEÆ. PENTANDRIA TRIGYNIA.

Named Maurocenia by Linnæus, in honour of Franc. Morosini, the Venetian senator; who had a fine garden at Padua.

THIS shrub bears a white blossom, which opens in July and August, and is succeeded by a fruit of a deep purple colour, from which the plant takes its familiar name. This shrub retains its leaves all the year; they are crisp and of a fine green, and when full of fruit, the plant is extremely handsome. Being a native of the Cape of Good Hope, it will not bear our winters abroad, but should be housed towards the end of September, and placed abroad again towards the middle or end of May. It must be sparingly watered; once a week in winter, but in dry summer weather three times.

HOUSELEEK.

SEMPERVIVUM.

CRASSULACEÆ. DODECANDRIA DODECAGYNIA.

French, joubarbe des toits [roof Jove's beard]; la grande joubarde [great Jove's beard]; jombarde; artichaut sauvage [wild artichoke]. —*Italian,* sempervivo.—*English,* Houseleek; Jupiter's-beard; Jupiter's-eye; Bullock's-eye; Sengreen; Aygreen; Live-ever; in the northern parts, Cyphel; perhaps from the Anglo-Saxon, Sinfulle.

THESE plants appear like a collection of large, glossy, green roses, of a heavy, leathery substance. Some persons admire and are very curious in them; others despise them as clumsy weeds. Linnæus informs us, that in Smoland, Houseleek is a preservative to the roofs of houses. The Common Houseleek may easily be made to cover the roof of a building, whether tile, thatch, or wood, by sticking the offsets upon it with a little earth.

The species vary in the colour of their flowers, and time of flowering; but they are most commonly red or yellow, appearing from June to August. The juice of the Houseleek, either alone or mixed with cream, affords immediate relief in burns and other external inflammations; and is considered an excellent remedy for the heat and roughness of the skin, sometimes attendant upon the changes of the seasons.

The most hardy kinds are the Common Houseleek, which is a native of most parts of Europe; the Globular, the Starry, the Cobweb, the Rough, the Mountain, and the Stone-crop-leaved. These will all thrive in the open air, and increase fast by offsets. They love a dry soil, and will spread very fast upon rocks or walls. A head dies soon after it has flowered; but it is soon supplied by offsets. If the common sort be planted in a little earth, upon a building or an old wall, it will thrive without any further attention. They are very succulent, and when planted in pots, must be very seldom, and very sparingly watered.

The Canary Houseleek must be housed in winter, admitting fresh air in mild weather; in the summer it must be so placed as to enjoy the morning sun.

The other kinds, with few exceptions, require the protection of a stove in the winter.

HYACINTH.

HYACINTHUS.

ASPHODELEÆ. HEXANDRIA MONOGYNIA.

Fabled to have sprung from the blood of Hyacinthus, when he was accidentally slain by Apollo with a quoit. Some derive the name from the Greek name of the violet, ια, and Cynthus, one of the names of Apollo.—*French,* Jacinthe des fleuristes; [Florist's Hyacinth]; Jacinthe Orientale [Oriental Hyacinth].—*Italian,* il giacinto; diacinto.

HYACINTHS may be blown either in earth or water; if in water, they may be set in the glasses any time between October and March, and by setting several in succession, may be continued for several months. The water should come a little above the neck of the glass, so that the bottom of the bulb may just sink below the surface. It will be well to place it in a part of the room where the sun can reach it, and it should have as much air as can conveniently be admitted into an inhabited room. The bulb will soon send out strong fibres below, and the stem will shoot above: these fibres form no mean portion of the beauty of the Hyacinth, and plead for its being placed in water rather than in earth. After it has begun to shoot, the water should be changed once a week, and before the stem is bent by the weight of the flower, it should be tied with a bit of green worsted to a stick, which some of the bulb-glasses are purposely made to admit.

Some persons have an earthenware vessel with a cover perforated with holes to admit the bottom of the bulbs: this being filled with water up to the cover, and a bulb placed upon each hole, with the bottom just dipping into the water, a number of flowers may be blown together, which will make a handsome display. The beauty of the fibrous roots is here as entirely lost as if the bulbs were planted in earth, an objection which would be obviated by using a vessel of glass in preference to one of earthenware, and their beauty would be seen to even greater advantage. Some persons put a little nitre in the water, which is said to improve the brightness of the colours in the flower.

When the bulbs are planted in earth, it should be done between November and February, and they should be placed within sight of the sun. A soil proper for them may be obtained from a nursery-man in the habit of furnishing plants, &c. The pot should be about seven inches in depth; the crown of the bulb about an inch and a half, or two inches deep, according to the size. When the plant begins to appear, the earth should be gently watered twice or thrice a week, as may be requisite to keep it rather moist. Fresh air may be admitted when convenient, as directed for those in water.

The flowers will blow in a shorter or longer time from

planting, in proportion to the warmth of their situation: such as are designed for later flowering, may be gradually accustomed to the open air in April; but when so treated, must be raised in a room without fire, or the change may be too great. These must enjoy the morning and evening sun, but be screened from the scorching heat of noon. They must likewise be sheltered from heavy rains; gentle showers will not be prejudicial to them; and such plants as are abroad will not require any other watering, except in a long continuance of dry hot weather.

When the plants have ceased flowering, the stalks, leaves, and fibres will decay; the principle of vegetation in the bulbs will be for a short time dormant: they should then be taken up, and laid in the open air, and in the shade, to dry. After a few days they may be removed into an airy room, and having remained uncovered until they are moderately hardened, the decayed parts, the loose skin, earth, &c. should be cleaned away, the offsets taken off, and the bulbs be put into a basket, or some dry place, where they may be secure from mice, &c. They may be preserved in this manner until it is time to replant them, when the old ones may be planted as before, and the offsets, two, three, or more in a pot, according to their size. Should any of the bulbs put out fibres while out of the earth or water, it will be necessary to plant or set them immediately, or they will be weakened, and will not flower with vigour. The leaves should never be plucked until they decay, or the bulb will thereby be deprived of a large portion of its proper nourishment.

The Grape Hyacinth, sometimes called Grape-Flower, is hardy, and will thrive in the open air. It is a native of the south of Europe, and blows in April or May. The flowers are blue, purple, white, or ash-coloured, and have an agreeable scent. The Purple Grape Hyacinth is called Tassel-Hyacinth. "The whole stalk," says Parkinson, "with the flowers upon it, doth somewhat resemble a long purse-tassel, and thereupon divers gentlewomen have so named it."—French, *Jacinthe à toupet* [Tufted Hyacinth]; *le vacinet de pres* [meadow myrtle]; in Lorraine, *ail de loup* [wolf's garlic]; in Anjou, *ail de chien* [dog's garlic]; *poireau bâtard* [bastard leek]; at Rochelle, *oignon sauvage* [wild onion]; *herbe du serpent* [snake wort]; in Provence, *lou congnou* [wolf's onion].—Italian, *cipolle canine* [dog onion].

But so many beautiful varieties have been raised from seed by attentive culture of the Eastern Hyacinth, that all the other species are comparatively neglected. The Eastern, or Garden Hyacinth, is a native of the Levant: it grows in abundance about Aleppo and Bagdad, where it flowers in February. With us, when not forced, it usually flowers in March or April. In Russia it has been found with yellow flowers. Culture has produced very large and double varieties of this Hyacinth. It is very sweet-scented, and much valued for the variety of its colours, which makes a number of them together appear very magnificent.

Mr. Miller says that we had formerly no other varieties of this Hyacinth in the English gardens than the single and double white, and blue: from the seeds of these, a few others were raised in England, and also by the Flemish gardeners, who came over annually with their flower-roots to vend in England: but the gardeners in Holland, within the last fifty years before his time, raised so many fine varieties, as to render the former of little or no value. Long after the Hyacinth had attracted the attention of florists, the double flowers seem to have been held in little esteem.

Peter Foorlem, of Haarlem, a noted cultivator of Hyacinths, was accustomed to throw the double flowers out of his collection; till once that he had been prevented by illness from visiting his flowers till just as they were going off, there happened to be one double flower remaining. It fixed his attention; not for any superior excellence, for it was small, and not particularly handsome; but, perhaps, because it was alone. He cultivated it, and increased it by offsets; florists saw, admired, and offered him a good price for it. He then became as zealous in cultivating his double flowers, as he had hitherto been in casting them away. The first double flower he raised was named Mary; this variety, and the next two that were produced, have been lost. The King of Great Britain is now considered as the oldest double Hyacinth. When it first appeared, it was preferred above all the other varieties then known; and the price of it was considerably above one thousand guilders. From that time great attention has been paid to the culture of this beautiful flower: and such has been the rage for it, that from one to two thousand guilders have been given for a single root. That is, from one to two hundred pounds sterling.

"The Haarlem gardeners have nearly two thousand varieties of the Hyacinth," says Mr. Martyn, "of which they generally publish a catalogue every year. New ones are annually produced, and in the circuit of that town alone, whole acres together are covered with these flowers."

The common Hyacinth, in French, *Jacinthe des bois* [Wood Hyacinth], is a native of Persia, and of many parts of Europe. In the spring it is very common in our woods, hedges, &c. and on this account, our old botanists have given it the name of the English Hyacinth. It is familiarly called the Harebell:

"In the lone copse, or shadowy dell,
Wild clustered knots of harebells blow:"
MRS. C. SMITH.

"The harebell for her stainless azured hue,
Claims to be worn by none but those are true."
W. BROWNE.

The fresh roots of this plant are said by Dr. Withering to be poisonous. Gerarde tells us that the juice which they contain answers the purposes of gum, and that with the exception of the Wake-Robin, it makes the best starch. This gum was used by fletchers to fix the feathers to arrows. The Harebell is sometimes white, or flesh-coloured, but much more commonly blue, or violet-coloured.

"The fanciful term of Hyacinthus non-scriptus, by which it is botanically distinguished," says Mr. Martyn, "was applied to this plant by Dodoneus, because it has not the Ai on the petals, and therefore is not the poetical Hyacinth."

It is not, indeed, supposed to be the Hyacinth of the ancient poets; but a flower which has been celebrated by Milton and Shakspeare possesses a just claim to the epithet poetical. They have stamped immortality on the Hyacinth of modern times.

———— "With fairest flowers,
Whilst summer lasts, and I live here, Fidele,
I'll sweeten thy sad grave: thou shalt not lack
The flower that's like thy face, pale primrose, nor
The azured harebell, like thy veins; no, nor
The leaf of eglantine, whom not to slander,
Outsweeten'd not thy breath."

The true poetical Hyacinth of the ancients is supposed to be the Red Martagon Lily. Mr. Martyn observes that most of the Martagons are marked with many spots of a darker colour than the flower itself; which often so run together as to form the letters Ai—as the ancient Hyacinth is represented.

Our modern Hyacinth has celebrity enough to stand upon its own ground, and, though it bears the same name, needs not to usurp the birth-right of its elder brother, of whose origin we are told:

——————— " Apollo with unweeting hand,
Whilome did slay his dearly loved mate,
Young Hyacinth, the pride of Spartan land;
But then transformed him to a purple flower."

It is always of the *purple* Hyacinth the poets speak: the modern purple is a deep blue; the Roman purple more resembled a light crimson. The flower which we now call the Hyacinth is often of a blue-purple colour, but is very seldom seen at all approaching to a Roman purple. A flower fabled to have sprung from blood, we may naturally suppose to have been of a somewhat similar colour.

Virgil, in speaking of the Hyacinth, uses an epithet peculiarly applicable to the Martagon Lily:

" ——————— et ferrugineos Hyacinthos."

GEORGIC. 4.

And any one who is acquainted with the Martagon Lily will immediately recognise the kind of iron-red here described; although the flower is often of a bright crimson, and the spots nearly black.

The very different manner in which our English poets describe the colour of the Hyacinth, proves it to be a different flower: who can confound a Roman purple with sapphire?

——————— " Shaded Hyacinth, alway
Sapphire queen of the mid-May."

KEATS.

"Hyacinth, with sapphire bell
Curling backward."

HUNT'S MASK.

"Some deep empurpled as the Hyacinth,
Some as the Rubin laughing sweetly red,
Some like fair Emeraudes, not yet well ripened."

Here Spenser, speaking of the various colours of the grapes according to their ripeness, expressly distinguishes the purple of the Hyacinth from the red, which was very similar to the Roman purple.

It has been common to compare the Hyacinth to curls; and the curling of its petals is common to both flowers; though perhaps the modern Hyacinth, in its form, bears more resemblance to a cluster of hair-curls:

"——— and hyacinthine locks
Round from his parted forelock manly hung
Clustering."

Collins has the same simile in his Ode to Liberty:

"The youths whose locks divinely spreading,
Like vernal Hyacinths in sullen hue."

It occurs again in Sir Philip Sidney's Arcadia:

"It was the excellently fair Queen Helen, whose jacinth hair curled by nature, but intercurled by art, like a fine brook through golden sands, had a rope of fair pearl, which now hiding, now hidden by the hair, did, as it were, play at fast and loose each with other, mutually giving and receiving richness."

The Persian poet, Hafiz, also compares the dark Hyacinth to the locks of his mistress *. Lord Byron makes the same comparison, as also does Sir W. Jones re-

* See the notes to Moore's Lalla Rookh.

peatedly. Lord Byron says the idea is common to the Eastern as well as to the Grecian poets.

Allusions to the letters *ai*, supposed to be seen upon the ancient Hyacinth, are made by many of the poets. It requires but little assistance from the imagination to read them on the Martagon lily.

"Del languido giacinto, che nel grembo
Porta dipinto il suo dolore amaro."

——"The languid hyacinth, who wears
His bitter sorrows painted on his bosom."

Mr. Hunt, in his Calendar of Nature, after dwelling a little upon the question, whether the Martagon lily is the true Hyacinth, quotes a passage from Moschus, which he thus renders in English:

"Now tell your story, hyacinth; and show
Ai, ai, the more amidst your sanguine woe."

One of our modern poems, also, has an allusion to this circumstance:

"While I with grateful heart gather him yellow
Daffodils, pinks, anemonies, musk-roses,
Or that red flower whose lips ejaculate
Woe,—and form them into wreaths and posies."
AMARYNTHUS.

The description of the Hyacinth in Ovid exactly answers to the Martagon lily:

"Sweet flower, said Phœbus, blasted in the prime
Of thy fair youth: thy wound presents my crime.
Thou art my grief and shame. This hand thy breath
Hath crush'd to air: I, author of thy death!
Yet what my fault? unless to have played with thee,
Or loved thee, (oh, too well!) offences be.
I would, sweet boy, that I for thee might die!
Or die with thee! but since the fates deny

So dear a wish, thou shalt with me abide:
And ever in my memory reside.
Our harp and verse thy praises shall resound;
And in thy flower my sorrow shall be found.

* * * * *

Behold! the blood which late the grass had dyed
Was now no blood: from whence a flower full blown
Far brighter than the Tyrian scarlet shone:
Which seemed the same, or did resemble right
The lily, changing but the red to white.
Nor so contented (for the youth received
That grace from Phœbus); in the flower he weaved
The sad impression of his sighs: which bears
Ai, ai, displayed in funeral characters."

SANDYS'S OVID, Book X.

There have been great disputes and differences about the Hyacinth: all were agreed that our modern Hyacinth was *not* the Hyacinth of the ancients; but the difficulty was to determine what *was*. The larkspur has laid claim to this honour, and some have supposed it to be the gladiolus, or corn-flag; but the best arguments have been urged in support of the Martagon lily, which is now pretty generally acknowledged to be the true heir to this ancient and illustrious race.

HYDRANGEA.

SAXIFRAGEÆ? DECANDRIA DIGYNIA.

The name is of Greek origin, and signifies a water-vessel.

ONE of the most common plants seen in our balconies, windows, &c. is the Shrubby Hydrangea. It is very handsome; not only for its great balls of blossom, but perhaps yet more for its large luxuriant leaves. It is a native of North America, and flowers in July, August, and September.

The Garden Hydrangea, or Chinese Guelder-rose, is a much smaller plant. The flower-balls of this are not larger than the European Guelder-rose. They are of a beautiful rose-colour, much deeper than the blossoms of the Shrubby Hydrangea, which are sometimes almost white, faintly tinged with pink; and sometimes, as faintly with blue. It is said that the blossom of this plant will take the colour of any thing by which it is shaded; but it is more probable that its colour is modified by soil, air, age, health, &c. The Garden Hydrangea is much valued for its profusion of bright rosy clusters. Its birth-place is unknown; but it is very commonly cultivated in the gardens of China and Japan. If placed where it may enjoy the air, and the light and warmth of the sun, this plant will flower better in a room than in the open air.

The Shrubby Hydrangea will live through our winters very well. If a severe frost destroys the stalks, the root will put out new ones in the spring: but if it is desired to preserve the stalks, that the plant may become larger, it will be safer to house it during the severity of the winter. It may be increased by parting the roots, which should be done late in October. When it is not intended to part the roots, and they have outgrown the pot they have been lodged in, this is the best time to remove them to a larger. The Hydrangea likes a rich soil, and is one of the most thirsty of plants: to which circumstance, no doubt, it owes its name. It is not, however, to be called intemperate, since its thirst is entirely constitutional; and it desires no richer draught than pure water. This must be given liberally. In winter, when there are no leaves to nourish, a small quantity thrice a week will be sufficient; but in the summer it must be lavishly watered every evening, and, if the weather be very hot, in the morning also. If this is neglected, the plant will droop with a kind of magical quickness; and a large draught of water will as suddenly revive it: but a frequent repetition of such changes would materially weaken the plant.

> "So have I often seen a purple flower,
> Fainting through heat, hang down her drooping head,
> But soon, refresh her with a gentle shower,
> Begin again her lively beauties spread,
> And with new pride her silken leaves display,
> And while the sun doth now more gently play,
> Lay out her swelling bosom to the smiling day."
>
> P. FLETCHER.

> "Like as a tender rose in open plain
> That with untimely drought nigh withered was,
> And hung the head; soon as few drops of rain
> Thereon distil and dew her dainty face
> Gins to look up, and with fresh wonted grace
> Dispreads the glory of her leaves gay."
>
> SPENSER.

Some of the gardeners have metamorphosed the word Hydrangea into *head-ranger*, as if it had been named after the chief officer of a park. The misnomers of plants are often as amusing as those of ships. The poplar has become a *popular* tree; the elm is called *ellum;* acacia is twisted into *casher;* nasturtium into *stertian;* the jonquil is termed *john-kill;* and the pyracantha, *pia-camphor*. Asparagus has so long been *sparrow-grass*, that it is now often deemed sufficient to term it *grass* alone. Loudon in his Encyclopædia of Gardening, lately published, has strongly recommended the formation of these misnomers to gardeners, as a means of enabling them to recollect the true names of plants: and has given some very amusing and ludicrous instances, as *cheese-monger* for casumunar, *Majocchi* for mioga.

HYPERICUM.

HYPERICEÆ. POLYADELPHIA POLYANDRIA.

French, la toute-saine [all-heal].—*Italian,* androsemo; erba rossa. *English,* Tutsan, evidently a corruption of the French name; and Park-leaves, because it is often found in parks.

THERE are a great number of Hypericums, all of easy culture. The following are the most generally cultivated in our gardens:

The Warted Hypericum. * The Olympian. The Shining. The Canary. The Chinese.	All these must be sheltered in winter.
* Great-flowered Hypericum, or Tutsan. Tall Hypericum. The Proliferous. Ascyron. Common Tutsan, or Hypericum.	All these are very hardy.

Those marked with an asterisk may be readily increased by parting the roots, in September or October. The flowers are yellow, generally in bloom from July to September; but the Chinese species will continue in blossom nearly all the year. The earth should be kept moderately moist.

The leaves of the Common Tutsan were formerly applied to fresh wounds, whence it obtained the French name, *la toute-saine,* and our name, Tutsan. It is a native of this country, and most other parts of Europe.

The Perforated Hypericum, or St. John's Wort,—in French, *le millepertius* [the many-pierced]; *l'herbe de St. Jean; le trucheran; le trescalan jaune:* in Italian, *pilatro;*

iperico ; *perforata*—is common in woods, hedges, &c. in almost every part of Europe. On account of its balsamic qualities it is useful in medicine: an infusion of it is made in the manner of tea; and the leaves given in substance are said to destroy worms. An infusion of the flowers and young tops of this plant in oil is used externally in wounds, &c. The flowers tinge spirits and oils with a fine purple colour; and the dried plant boiled with alum dyes wool yellow.

Mistaking the meaning of some of the medical writers, who, from a supposition of its utility in hypochondriacal disorders, have given it the fanciful name of *fuga dæmonum* [devil's flight], the common people in France and Germany gather it with great ceremony on St. John's-day, and hang it in their windows as a charm against storms, thunder, and evil spirits. In Scotland, also, it is carried about as a charm against witchcraft and enchantment; and they fancy it cures ropy milk, which they suppose to be under some malignant influence. As the flowers, rubbed between the fingers, yield a red juice, it has also obtained the name of *sanguis hominis* [human blood] among fanciful medical writers.

Cowper speaks of the Hypericum as remarkably full of blossom: the species vary in this particular:

> " Hypericum all bloom, so thick a swarm
> Of flowers, like flies clothing her slender rods,
> That scarce a leaf appears."

INDIAN CORN.

ZEA.

GRAMINEÆ. MONOECIA TRIANDRIA.

Called also Maize; Turkey Corn.—*French,* le mais; mayz; blé de Turquie; blé d'Espagne [Spanish corn]; blé de Guinée [Guinea corn]; ble d'Inde [Indian corn]; gros millet des Indes [great Indian millet].—*Italian,* gran Turco, furmento Turco [Turkey wheat]; formentone; grano d'India.

THIS corn should be sown early in April, in large deep pots. It may be sown, at first, several in one pot, and afterwards removed; transplanting them into separate pots about the end of May. It will not grow so high in a pot as in the open ground, but is worth raising in this manner for the sake of its long elegant leaves. It should stand in the open air, and, in dry weather, be watered every evening. If there is convenient room for it in-doors, the seed may be sown a month earlier, and kept under cover till the beginning or middle of April. The plant will decay in the autumn.

INDIAN PINK.

DIANTHUS CHINENSIS.

Called also China Pink.—*French,* l'oeillet de la Chine.

THE Indian Pink is generally considered as an annual plant, and therefore the roots are not often preserved; but, if they are planted in a dry soil, they will often produce finer flowers the second year than the first, and in greater number. It is a very ornamental plant, from the various and beautiful colours of its blossoms. It may be sown early in April; if in a pot six inches wide, only one; but they look better sown in a box, many together, about six inches apart. They may stand abroad: in dry weather they should be watered three times a week. They will flower from July till the approach of frost; if they are then cut down, the root will generally put out new stalks, and flower well the next year.

IPOMŒA.

CONVOLVULACEÆ. PENTANDRIA MONOGYNIA.

This genus is very nearly allied to the Convolvulus, and the name is derived from Greek words, expressive of its similarity to that flower.

The Ipomœa is very beautiful, but unfortunately very tender, being chiefly Indian. One species in particular would be desirable. The Ipomœa Quamoclit—in French, *Jasmin rouge de l'Inde* [Red Indian Jasmine]; *Fleur de Cardinal* [Cardinal-flower].—Italian, *Quamoclito*—which is the most beautiful of them all, in colour and in form, in leaf and in flower. "It is a beautiful climber," says Sir W. Jones; "its blossoms are remarkably elegant and of a rosy red." It has the scent of cloves. It is called by the Indians Camalata, or Love's-creeper.

There are two kinds which may be procured at a nursery, and preserved through the summer in an inhabited room: the Coccinea, or Scarlet-flowered Ipomœa, and the Nightshade-leaved, the blossoms of which are of a pale rose colour. The earth must be kept moderately moist, but water must be given but in small quantities at one time. The plant will require support.

There is a species of the Ipomœa, which, from one root, may be carried over an arbour three hundred feet in length: it is a perennial species, and is called in Jamaica, the Seven-year-vine, or Spanish Arbour-vine.

"The Camalata," says Sir W. Jones, "is the most lovely of its order, both in the colour and form of its flowers; its elegant blossoms are 'celestial rosy red, Love's proper hue,' and have justly procured it the name of Camalata, or Love's-creeper. Camalata may also mean a mythological plant, by which all desires are granted to such as inherit the heaven of India; and if ever flower was worthy of Paradise, it is our charming Ipomœa."

IRIS.

IRIDEÆ. TRIANDRIA MONOGYNIA.

So named for its variety of colours. It is also named Flower-de-luce.—*French*, fleur-de-lys; iris; flambe; and glaieule; in the village dialect, glè; baguettes.—*Italian*, iride; giaggiolo; giglio.

THIS flower claims the whole world as her country: some few species are from America; several are natives of the colder regions of Asia; still more, of Europe; and most of all, of the Cape of Good Hope.

Some of the species have very large flowers, which, from their colours being very vivid, and several uniting in the same blossom, are extremely showy. Many of them are bulbous-rooted; and of these the most esteemed is the Persian Iris; for the beauty and fragrance of its flowers, and for their early appearance; for it is generally in full perfection in February, or early in March. A few of these flowers will perfume a whole room: their colours are a mixture of pale sky blue, purple, yellow, and sometimes white. This kind, the Tuberous-dwarf, and the Spanish bulbous Iris, may be blown in water-glasses, as directed for the Hyacinth; only that these will not so well bear to be reserved for late planting, because they are apt to shrink. It is better on this account to plant them in October, or soon after; at any rate by the end of the year. But they are thought to flower stronger in pots: they like a light sandy earth, and take delight in the morning sun: the more fresh air they have allowed them, when not frosty, the better they will thrive. The earth should be kept always moist; and, when the flowers and leaves have decayed, the bulbs should be treated as those of the Hyacinth. After they have lain out of the ground about a month, they should be frequently examined; for if they begin to shrink, they must be planted immediately. The bulbs should be set two inches deep, from the surface of the earth to the top of the crown.

The Chalcedonian Iris—in French, *l'iris de Suse*, or *de Constantinople*—has also very large flowers, and is the most magnificent of them all; but the petals are very thin, and hang in a kind of slatternly manner, which makes it appear, to some persons, less handsome than others which are smaller. It likes a loamy earth, and sunny exposure; this species must be very sparingly watered: moisture favours the growth of all the other kinds; but it will injure this to give it more than will preserve the earth from absolute drought. It may be increased by parting the roots in autumn; and during the winter months it will be safer to give it house-room. This flower is called, by old writers, the Turkey Flower-de-luce.

The Snakes-head Iris may be increased in the same manner, but must be kept moist, and needs no winter shelter: the flower is dark purple, approaching to black.

The Twice-flowering Iris—in French, *l'iris des deux saisons*—(so called because it flowers both in spring and autumn) the Various-coloured, the Pale-yellow, the Grass-leaved, and the Siberian, may be treated as the Snakes-head species: they like an eastern aspect.

To these may be added the Florentine Iris, White Fleur-de-luce, or Flower-de-luce of Florence—in French, *la flambe blanche*—and the Crocus-rooted Iris, or Spanish Nut: these should be housed in winter, but merely to protect them from frost. The root of the former, corruptly called orrice, is used to communicate a violet scent to hair-powder, oils, and syrup.

But it is useless to enumerate more: these already named are the best adapted for pots, and the Persian Iris is far preferable to all the others. Irises may be removed every year, or second year, as most convenient. The Persian may be kept till the third.

The Common Yellow Iris is called Water-flag; Yellow-flag; Water-sedge; in Scotland, Water-skeggs; Lugs—in French, *le glayeul des marais* [marsh-flag]; *la flambe bâtarde* [false iris]; *le faux acore* [false acorus]; *la flambe aquatique* [water iris]; *le glayeul à fleur jaune* [yellow flag]:—Italian, *iride gialla.*—This and several other kinds have valuable medicinal properties: the root may be used instead of galls, in making ink, or black dye. The seeds are the best substitute for coffee, hitherto discovered. The juice is sometimes used as a cosmetic for removing freckles, &c., and a most beautiful colour for painting has been prepared from the flowers.

" Many of the African kinds," says Mr. Martyn, " are eaten both by men and monkeys; and the roots, when boiled, are esteemed pleasant and nourishing."

Although the Iris is not considered as a Lily, the French have given it the name of one; it is the Fleur-de-lys, which figures in the arms of France. The Abbe la

Pluche, in *La Spectacle de la Nature,* gives the following conjectural origin of this name:

"The upper part of one leaf of the Lily, when fully expanded, and the two contiguous leaves beheld in profile, have," he observes, "a faint likeness to the top of the Flower-de-luce: so that the original Flower-de-luce, which often appears on the crowns and sceptres in the monuments of the first and second race of kings, was most probably a composition of these three leaves. Lewis the Seventh engaged in the second crusade; distinguished himself, as was customary in those times, by a particular blazon, and took this figure for his coat of arms; and as the common people generally contracted the name of Lewis into Luce, it is natural," says the Abbe, "to imagine that this flower was, by corruption, distinguished in process of time by the name of Flower-de-luce." But some antiquaries are of opinion, that the original arms of the Franks were three toads; which, becoming odious, were gradually changed, so as to have no positive resemblance of any natural object, and named Fleur-de-lys.

Shakspeare appears to consider this flower as a Lily only by courtesy:

> ————————"lilies of all kinds,
> The Flower-de-luce being one."

G. Fletcher gives a pretty picture of this flower:

> "The Flowers-de-Luce, and the round sparks of dew
> That hung upon their azure leaves did shew
> Like twinkling stars, that sparkle in the evening blue."

Drayton expressly distinguishes the Flower-de-luce from the Lily.

> "The Lily, and the Fleur-de-lis,
> For colour much contenting!
> For that I them do only prize,
> They are but poor in scenting."

The poet seems not to have been acquainted with the Persian Iris, which has so fine a perfume.

Spenser also distinguishes the Flower-de-luce from the Lily, though acknowledging the connexion:

> "The Lilly, lady of the flowering field,
> The Flower-de-luce, her lovely paramour,
> Bid thee to them thy fruitless labours yield,
> And soon leave off this toilsome weary stour:
> Lo! lo! how brave she decks her bounteous bower
> With silken curtains, and gold coverlets,
> Therein to shroud her sumptuous balamour;
> Yet neither spins, nor card, ne cares, nor frets,
> But to her mother Nature all her cares she lets."

Mrs. C. Smith gives a lively picture of the Yellow, or Water Iris.

> "Amid its waving swords, in flaming gold
> The Iris towers ———."

IXIA.

GLADIOLEÆ. TRIANDRIA MONOGYNIA.

THERE are many species of Ixia, varying in colour: they have bulbous roots, and may be increased by their offsets; but they will not flower well if parted oftener than every third year. In the autumn the stalks and leaves decay; the roots should then be put under shelter for the winter, unless it is designed to remove them; in which case they may be treated in the same manner as the Hyacinth, and bulbs in general, and may be replanted any time between October and January. They may stand abroad in the summer, and should then have a little water every evening: they should be sparingly watered in winter, when left in the earth. Pots three inches in diameter, and five in depth, will be large enough for these plants: the bulbs should be covered about an inch deep.

JERUSALEM-SAGE.

PHLOMIS.

LABIATÆ. DIDYNAMIA GYMNOSPERMIA.

Called also Tree-sage.—*French,* bouillon blanc de Sicile [white mullein of Sicily]; sauge en arbrisseau; sauge en arbre.

THIS shrub retaining its leaves all the year, and its bright yellow flowers the greater part of the summer, is very desirable. It should be sheltered from severe frost; but in mild winters, if not convenient to house it, a little saw-dust laid over the roots will be a sufficient protection. In dry summer weather it may be allowed a little water every evening: once or twice a week will suffice in the winter, and none during frost.

JESSAMINE.

JASMINUM.

JASMINEÆ. DIANDRIA MONOGYNIA.

The name of this plant is derived from the Greek, and signifies an agreeable odour. Nearly all the European languages have the same name for it.—In *French,* it is jasmin: in *Italian,* gelsomino: *Spanish,* jasmin: *Dutch,* jasmyn, &c. &c. In *English* it is sometimes familiarly called Jessamy, Jessima, and Gesse.

THE kinds of Jessamine most frequently grown in pots are the Yellow Indian, and the Spanish or Catalonian.

The first grows to the height of eight or ten feet; the leaves continue green all the year, and the blossoms are of a bright yellow, very fragrant, and blowing from July till October or November. They are frequently succeeded by oblong berries, which turn black when ripe.

The Spanish Jessamine, so named because it came to us from Spain, is a native of the East Indies. The flowers are of a blush-red outside, and white within; blowing at the same time as the Indian kind. From the middle of May to the middle of October they may stand in the open air; but must then be housed, having as much fresh air as possible in mild weather. They should have but little water at a time, but that should be given often, so that the earth may be always moist. In spring, the decayed branches should be pruned; and of the Spanish kind the sound ones should be pruned to the length of two feet, which will cause them to shoot strong, and produce many flowers. But this liberty must not be taken with the Indian kind.

The Common White Jessamine is an exceedingly elegant plant for training over a wall, where that support can be allowed, and after its first infancy will bear our winters very well. It is a delicate and fragrant shrub, not surpassed in beauty by any of the species. It is of this Cowper speaks in the following passage:

> " The jasmine throwing wide her elegant sweets,
> The deep dark green of whose unvarnished leaf
> Makes more conspicuous, and illumines more,
> The bright profusion of her scattered stars."

The Hindoos, who use odoriferous flowers in their sacrifices, particularly value the Jessamine for this purpose, and the flower which they call Zambuk.

Jessamine is one of the shrubs of which Milton forms the bower of Adam and Eve in Paradise:

"Thus talking, hand in hand alone they pass'd
On to their blissful bower: it was a place
Chosen by the sovereign Planter, when he framed
All things to man's delightful use; the roof
Of thickest covert was inwoven shade,
Laurel and myrtle, and what higher grew
Of firm and fragrant leaf; on either side
Acanthus, and each odorous bushy shrub,
Fenced up the verdant wall; each beauteous flower,
Iris all hues, roses, and jessamine,
Rear'd high their flourish'd heads between, and wrought
Mosaic; underfoot the violet,
Crocus, and hyacinth, with rich inlay
Broider'd the ground, more colour'd than with stone
Of costliest emblem."

Mr. T. Moore speaks of the Jessamine as more fragrant by night than by day:

"'Twas midnight—through the lattice, wreathed
With woodbine, many a perfume breathed
From plants that wake when others sleep;
From timid jasmine buds, that keep
Their odour to themselves all day,
But, when the sun-light dies away,
Let the delicious secret out
To every breeze that roams about."

"The jessamine, with which the queen of flowers,
To charm her god, adorns his favourite bowers;
Which brides by the plain hand of Neatness drest,
Unenvied rival! wear upon their breast;
Sweet as the incense of the morn, and chaste
As the pure zone which circles Dian's waist."

CHURCHILL.

Jessamine abounds in Italian gardens. In the East it is cultivated for the stems, of which pipes are made.

JUNIPER.

JUNIPERUS.

CONIFERÆ. DIOECIA MONADELPHIA.

French, le génevrier; le genièvre; le petron: in the old writers, jupicelle; genibretos; cadenelo: in Languedoc, lou geniebre: in Provence, genibre.—*Italian,* il ginepro: in the Brescian, zenéver: at Venice, brusichio.

THE Common Juniper is well adapted for potting, and is the more desirable as being an evergreen. It is common in all the northern parts of Europe, in rich or barren soils, in open sandy plains, or in moist close woods: it will bear the severest cold in our climate, and will require no other attention than to keep it clear from weeds, and to give it a little water in a continued drought.

This shrub is celebrated for its medicinal properties: a sweet decoction is made from the berries, from which a quantity of sugar may be obtained. The bark may be made into ropes. A spirit impregnated with the essential oil of the berries is known by the name of Holland gin, or Hollands. The common English gin also derives part of its flavour from these berries, but is a very compound liquor. We are told by Linnæus that the Swedes prepare a beer from them, which they consider very efficacious in scorbutic cases; and that, for the same purpose, the Laplanders drink an infusion of them, as we do tea or coffee. Juniper wine is sometimes made, and is said to be a very wholesome one.

Who would suppose that the word Gin has most likely a common origin with a female name famous in poetry and romance—Ginevra, or Gineura? The Italian word for Juniper is Ginebro, or Ginepro, which, by an alteration common to the South, becomes Ginevro. The French word is Genevre, corrupted into our word Geneva. The name of Ariosto's favourite lady was Gineura, which gave him occasion to immortalize the Juniper-tree, as Petrarch did the laurel. He says, in one of his sonnets, that with the leave of Apollo and Bacchus he will be crowned with Juniper, and not with the bay or the ivy:

"Quell' arboscel, che in le solinghe rive
A l' aria spiega i rami orridi ed irti,
Ed' odor vince i pin, gli abeti, e i mirti,
E lieto e verde al caldo, e al ghiaccio vive.
Il nome ha di colei, che mi prescrive
Termine e leggi a' travagliati spirti
Da cui seguir non potran Scille o Sirti,
Ritrarmi, o le brumali ore, o lo estive:
E se benigno influsso di pianeta
Lunghe vigilie, od amorosi sproni
Son per condurmi ad onorata meta;
Non voglio (e Febo, e Bacco mi perdoni)
Che lor frondi mi mostrino poeta,
Ma che un Ginebro sia che mi coroni."

ARIOSTO, SONN. 7.

"The shrub that on solitary shores spreads to the air its dark and bristled branches, outscenting pines, and firs, and myrtles; still green in summer's heat, and winter's cold; bears the name of her who prescribes terms and laws to my troubled soul: laws that I will not be turned from following, either by rocks, or whirlpools; either in the wintry season, or the summer. And if the benign influences of the planet, long watchings, or amorous zeal, are to conduct me to that height of honour, I will not (Phœbus and Bacchus pardon me) that their leaves should declare me for a poet, but that a Juniper should crown my brow."

Tasso, in his miscellaneous poems, has two sonnets to a similar purpose.

Before the use of carpets in Europe, the richest people used to strew their apartments with dried leaves and rushes. Queen Elizabeth walked on no better floor. The gentlemen and ladies in Boccaccio are luxurious enough to walk on flowers of Juniper. "This jocund company," says an old translation, "having received licence from their queen to disport themselves, the gentlemen walked with the ladies into a goodly garden, making chaplets and nosegays of divers flowers, and singing silently to themselves. When they had spent the time limited by the queen, they returned into the house, where they found that Parmeno had effectually executed his office; for when they entered into the hall, they saw the tables covered with delicate white napery, and the glasses looking like silver, they were so transparently clear;—all the room besides strewed with flowers of Juniper."

As the passage has to do with gardens and flowers, and is a very elegant one besides, the reader will not object to a quotation of the whole of it:

"When the queen and all the rest had washed, according as Parmeno gave order, so every one was seated at the table: the viands, delicately dressed, were served in, and excellent wines plentifully delivered: none attending but the three servants, and little or no loud table-talk passing among them. Dinner being ended, and the table withdrawn, all the ladies, and the gentlemen likewise, being skilful both in singing and dancing, and playing on instruments artificially, the queen commanded that divers instruments should be brought; and as she gave charge, Dioneus took a lute, and Fiametta a viol-de-gamba, and began to play an excellent dance: whereupon the queen, with the rest of the ladies, and the other two young gentlemen (having sent their attending servants to dinner), paced forth a dance very majestically, and when the dance was ended, sung sundry excellent canzonets, outwearing so the time until Parmeno commanded them all to rest, because the hour did necessarily require it. The gentlemen having their chambers severed from the ladies, curiously strewed with flowers, and their beds adorned in exquisite manner, as those of the ladies were not a jot inferior to them. The silence of the night bestowed sweet rest on them all. In the morning, the queen and all the rest being risen, accounting overmuch sleep to be very hurtful, they walked abroad into a goodly meadow, where the grass grew verdantly, and the beams of the sun heated not over violently, because the shades of fair-spreading trees gave a temperate calmness, cool and gentle winds fanning their sweet breath pleasingly among them." The company then sit down, and the celebrated novels commence.

It is still a common custom in Sweden to strew the floors with sprigs of Juniper*.

KALMIA.

RHODORACEÆ. DECANDRIA MONOGYNIA.

So named by Linnæus in honour of Peter Kalm, professor at Abo in Sweden.

The Kalmias are handsome shrubs, bearing flowers in clusters, of a rose or peach colour. The Broad-leaved species grows much higher than the others: they must be obtained from a nursery, and will require to be watered pretty liberally. They are natives of North America.

* See Clarke's Travels, vol. iii.

LARKSPUR.

DELPHINIUM.

RANUNCULACEÆ. POLYANDRIA TRIGYNIA.

The Latin name was given to this plant from an idea that the buds had some resemblance to a dolphin.—*French,* pied-d'alouette; l'eperon de chevalier [knight's spur]; la consoude royale [royal comfrey]; l'herbe Sainte Othilie.—*Italian,* speronella [little spur]; sperone di cavaliere; consolida reale; fior regio [king flower].—*English,* larkspur; lark's-claws; lark's-heel; lark's-toe, on account of the spur-shaped nectary at the back of the flower.

THE Branching or Wild Larkspur grows naturally in many parts of Europe: it varies in colour. From the flowers, when blue, a good ink has been made, with the addition of a little alum.

All the Larkspurs are hardy, and may be easily raised from seed; but as the perennial kinds do not flower the first year, it is better to procure them from a nursery. The annual kinds, by sowing in succession in September, October, March, and April, may be had in blossom from the beginning of June to the end of September: one seed in a pot of at least eight inches. They do not well bear transplanting.

Those sown in the autumn will produce the strongest flowers. They should stand abroad, and in dry summer weather be watered a little every evening; but water must be given sparingly in the winter.

Linnæus and some others are of opinion that the Larkspur is the hyacinth of the poets; but this opinion is considered as unfounded. Professor Martyn has determined the Martagon lily to be the ancient hyacinth, and the learned Heyne coincides with him. (See Hyacinth.)

LAUREL.

PRUNUS LAURO-CERASUS.

ROSACEÆ. ICOSANDRIA MONOGYNIA.

French, le laurier-cerise.—*Italian,* lauro regio; lauro di Trebisonda.

THE Laurel, which has been frequently confounded with the Laurus Nobilis, or Sweet Bay, does not even belong to the same genus. Among the species of Laurus are many valŭable trees, as the camphor, sassafras, cassia, cinnamon, &c.; but the common Laurel is not one of that family.

This Laurel was formerly called the Cherry-bay, or Bay-cherry, and was preserved in green-houses in the winter. The only protection against the climate now afforded it, is in planting it in a warm aspect, or against a warm wall, to preserve it from frost. In warm countries, the Laurel will grow to a great size; so that in some parts of Italy there are large woods of them. Where they are numerous, and near together, they defend each other, and are not liable to injury by frost: but when in pots, the roots should be covered with a little straw in severe winters. In dry winter weather, when not frosty, it may be watered once a week; in the summer, every evening when there is no rain.

The Portugal is much hardier than the Common Laurel. They may be increased by cuttings of the same year's shoots, which should be planted in September. If a small part of the former year's wood be left at the bottom, they will root faster. They should be planted five or six inches deep, in a soft, loamy earth, and the earth pressed close to them.

Evelyn says, that if the Lauro-cerasus, or Cherry-laurel, were not always suffered to run so low and shrubby, it would make a handsome tree on a stem, with a head resembling the orange. The way to have this tree of a handsome shape, with an upright stem, and the boughs regularly disposed, is to raise it from the berry. This is also the case with the bay, the orange. and many others.

LAVATERA.

MALVACEÆ. MONADELPHIA POLYANDRIA.

So named from Lavater, a physician at Zurich.

The Lavateras are large, handsome flowers, in form resembling the mallow, but considerably larger. In colour they vary from a pure white to blue, flesh-colour, &c.

The annual kinds should be sown in autumn, in small pots; one in each. Towards the end of October they should be removed into the house, and, being defended from frost, will abide the winter very well. Early in April they may be shaken out of their pcts, and planted in larger; where they may remain to flower, which they will do in July.

The perennial kinds are not so well adapted for potting; and the annual will furnish a sufficient variety of these elegant flowers for any house, balcony, &c. In dry summer weather they may be gently watered every evening. If perennial kinds are obtained, they must be sheltered in the winter, and be sparingly watered in that season. Most of the species require sticks to support them.

LAVENDER.

LAVANDULA.

LABIATÆ. DIDYNAMIA GYMNOSPERMIA.

So named from its use in fomentations and baths.—*French,* la lavande: in Provence, aspic; espic, whence the foreign oil of lavender is usually called oil of spike.—*Italian,* lavendola; lavanda; spigo.

The Common Lavender is increased by cuttings, which should be planted in March, and placed in the shade until they have taken root: they may then be exposed to the sun. These plants will live much longer, and endure the most severe cold, if planted in a dry, gravelly soil. They grow faster in summer if the soil be rich and moist; but then they are generally destroyed in the winter, nor are they so strongly scented as those which grow in a barren soil.

Lavender was formerly used for edgings, as we now use box, thrift, &c.; but it grows too high for this purpose, and the practice is generally discontinued. The agreeable scent of Lavender is well known, since it is an old and still a common custom to scatter the flowers over linen, as some do rose-leaves, for the sake of this sweet odour:

> "Pure lavender, to lay in bridal gown."

Lavender-water, too, as it is usually called, although it is really spirit of wine scented with the oil of lavender, is one of our most common perfumes. This plant has been much celebrated for its virtues in nervous disorders, and is an ingredient in some of the English-herb teas now in such general use. This species of Lavender is common to Europe, Asia, and Africa. It flowers from July to September.

French Lavender (also called Purple Stoechas, from being found in the islands named the Stoechades) may be sown in March; several seeds together in a light, dry soil. When the plants are two inches high, they may be separated, and planted into pots seven inches wide: they must be placed in the shade till they have taken root, and be gently watered every second day. If the winter prove severe, they should be housed; but in a dry soil they will bear our common winters very well. This species may also be raised from cuttings, like the Common Lavender. It is a native of the South of Europe, and is in bloom from May to July.

The other kinds of Lavender may be increased either by cuttings or seeds; but they do not all ripen seeds in this country. They require winter shelter; and the Thick-leaved species, which is a native of the East Indies, must be preserved in a hot-house.

They should have but just water enough to prevent drought; especially in the winter.

The stalks of the Lavender, even when the flowers have been stripped away, have an agreeable scent; and if burnt, will diffuse it powerfully and pleasantly: they form an agreeable substitute for pastiles, and will burn very well in the little vessels made for that purpose. To a Londoner it becomes a kind of rural pleasure to hear the cry of—"Three bunches a penny, sweet Lavender."

"And lavender, whose spikes of azure bloom
Shall be erewhile in arid bundles bound,
To lurk amidst her labours of the loom,
And crown her kerchiefs clean with mickle rare perfume."

SHENSTONE'S SCHOOL-MISTRESS.

LEMON-TREE.

CITRUS-LIMON.

AURANTIACEÆ. POLYADELPHIA POLYANDRIA.

French, le limonier; l' arbre du limon.—*Italian,* limone.

The Lemon may be treated like the Orange-tree; and as the treatment is given at great length under that head, it would be useless to repeat it. The only difference is, that the Lemon, being rather hardier, may be placed in the more airy part of the room in winter, and may have rather more water; though the orange must be frequently supplied, even in winter, unless it be a bitter frost.

The Lemon is a variety of the citron, which was first known in Europe by the name of the Median-apple, being brought from Media. Virgil terms it the "*happy* apple:" "probably," says Mr. Davidson, "on account of its great virtues:"

> "Media fert tristes succos, tardumque saporem
> Felicis mali; quo non præsentius ullum,
> Pocula si quando sævæ infecere novercæ,
> Miscueruntque herbas, et non innoxia verba,
> Auxilium venit, ac membris agit atra venena."
>
> GEORGIC 2.

> "Nor be the citron, Media's boast, unsung,
> Though harsh the juice, and lingering on the tongue:
> When the drugg'd bowl, mid witching curses brew'd,
> Wastes the pale youth by step-dame hate pursued,
> Its powerful aid unbinds the mutter'd spell,
> And frees the victim from the draught of hell."
>
> DR. PARIS'S TRANSLATION.

Martyn, in his Notes, cites a story related by Athenæus of the use of citrons against poisons, which he had from a friend of his, who was governor of Egypt. This governor had condemned two malefactors to death by the bite of serpents. As they were led to execution, a person, taking compassion on them, gave them a citron to eat. The consequence of this was, that though they were exposed to the bite of the most venomous serpents, they received no injury. The governor being surprised at this extraordinary event, inquired of the soldiers who guarded them, what they had eaten or drank that day, and being informed that they had only eaten a citron, he ordered that the next day one of them should eat citron, and the other not. He who had not tasted the citron died presently after he was bitten; the other remained unhurt.

LILAC.

SYRINGA.

JASMINEÆ. DIANDRIA MONOGYNIA.

French, lilas commun; lilas; queue de renard de jardin [garden fox-tail].—*Italian,* siringa: in Sicily, alberu di pacenzia.

THE name Syringa is of Greek origin, and signifies a pipe. The old English name is Pipe-tree. Caspar Bauhin supposes Syringa to be an African wood. Linnæus was inclined to trace the name to the nymph Syrinx, who, to escape the pursuit of the god Pan, was, at her own request, changed by the gods into a reed; of which Pan formed a musical instrument, and gave it the name of his favourite nymph:

> "Among the Hamadryade Nonacrines,
> (On cold Arcadian hills) for beauty famed,
> A Nais dwelt; the nymphs her Syrinx named,

Who oft deceived the satyrs that pursued,
The rural gods, and those whom woods include.
In exercises, and in chaste desire,
Diana-like; and such in her attire.
You either in each other might behold,
Save that her bow was horn—Diana's, gold:
Yet oft mistook. Pan, crown'd with pines, returning
From steep Lycæus, saw her; and love-burning,
Thus said: 'Fair virgin, grant a god's request,
And be his wife.' Surcease to tell the rest;
How from his prayers she fled, as from her shame,
Till to smooth Ladon's sandy banks she came:
There stopp'd; implored the liquid sister's aid
To change her shape, and pity a forced maid.
Pan, when he thought he had his Syrinx clasp'd
Between his arms, reeds for her body grasp'd.
He sighs: they, stirr'd therewith, report again
A mournful sound, like one that did complain.
Rapt with the music—'Yet, oh sweet!' said he,
'Together ever thus converse will we.'
Then of unequal wax-join'd reeds he framed
This seven-fold pipe: of her 'twas Syrinx named."

SANDYS' OVID, book 1.

Lilac, or Lilag, is a Persian word, signifying a flower.

Of the Common Lilac there are three varieties: the Blue, the Violet, and the White. The second is generally known by the name of the Scotch Lilac: this has the fullest flowers.

"The Lilac," says Mr. Martyn, "is very commonly seen in English gardens, where it has long been cultivated as a flowering shrub. It is supposed to grow naturally in some parts of Persia; but it is so hardy as to resist the greatest cold of this country.

"The Scotch Lilac," continues he, "is the most beautiful of the three; and is probably so called because it was first mentioned in the catalogue of the Edinburgh garden."

Gerarde and Parkinson cultivated the Blue and the White kinds under the name of Pipe-tree, or Pipe-privet. Gerarde says, "I have them growing in my garden in great plenty." (1597.) This shows it to have been at that time comparatively rare; and the beautiful Lilac now so common in our gardens and shrubberies was far more so.

The flowers appear towards the end of April, or early in May, and usually last about a month. Although called a shrub, the Lilac will grow to the height of eighteen or twenty feet; and the leaves growing very luxuriantly, it may be considered as a tree of very respectable dimensions.

The species of Lilac best adapted for pots is the Persian, which seldom exceeds six feet in height. The flowers blow some weeks later than those of the Common Lilac, and last longer in beauty; but do not produce ripe seeds in England. It is a light and elegant shrub, of a more lady-like delicacy than the Common kind; compared to which, it is as the light and crisp Chinese-rose compared to the full-blown beauty of the Cabbage-rose. This shrub was formerly known among the nurserymen by the name of the Persian Jasmin.

The Common Lilac thrives best upon a rich, light soil, such as the gardens in the neighbourhood of London are chiefly composed of; and there they grow much larger than in any other part of England. In a strong loam, or a chalky soil, they make little or no progress. The best time to transplant them is in the autumn. The Common Lilac is a native of Persia, as well as the Persian, specifically so called; but the latter was brought to this country about half a century later.

It may not be altogether useless to mention, that the flowers of the Lilac are always produced upon the shoots of the former year; and below the flowers, on the same shoot, other shoots come out to succeed them; for that part upon which the flowers stand decays down to the shoots below every winter. Therefore, if it is desired to preserve the tree in full beauty, care should be taken in plucking the flowers, not to take with them those young shoots which are to produce the flowers of the following season, or the blossoms will be comparatively few.

The earth should be kept moderately moist, and the Persian Lilac should be sheltered from frost.

When the Lilac blossom has attained its full beauty, it begins to fade gradually, until it becomes at last of a red colour. Thus Cowper speaks of them as sanguine:

"The lilac, various in array, now white,
Now sanguine, and her beauteous head now set
With purple spikes pyramidal, as if
Studious of ornament, yet unresolved
Which hues she most approved, she chose them all."

COWPER'S TASK.

——— "shrubs there are
Of bolder growth, that at the call of spring
Burst forth in blossom'd fragrance; lilacs robed
In snow-white innocence, or purple pride."

THOMSON'S SPRING.

LILIES.

LILIUM.

LILIACEÆ. HEXANDRIA MONOGYNIA.

French, le lis; lys.—*Italian,* giglio: in the Brescian, zei.

Although we usually associate the idea of extreme whiteness with the lily, so that it is common to express a pure white by comparison with this flower, as with snow, and as white as a lily is an old and common proverb, yet lilies are of almost every variety of colour: perhaps there is no other flower that varies so much in this respect.

"The Common White Lily," says Mr. Martyn, "has been cultivated in England time immemorial." The stem is usually about three feet high. The flowers are brilliantly white, and glossy on the inside. It is from the East; and in Japan the blossom is said to be nearly a span in length. This Lily flowers in June and July. The roots, which are mucilaginous, are sometimes boiled in milk or water, and employed in emollient poultices; but they have not much reputation. An oil for the same purpose was also prepared by infusing the roots in olive oil.

There are several varieties of the White Lily: as, that with the flowers striped or blotched with purple; that with the leaves striped or edged with yellow; one with double, and one with pendulous flowers. The double flowers are less fragrant than the single; and the common kind is generally held in higher estimation than any of the others.

This Lily may easily be increased by offsets, which the bulbs furnish in great plenty. They should be taken off every second year. The best time to remove it is about the end of August, soon after the stalks decay. It will thrive in almost any soil or situation, is very hardy, and not liable to injury by frost. Few plants are more easily increased or preserved than the Lily, so remarkable for the beauty and fragrance of its flowers.

The bulbs, when removed, may be treated as other bulbs; but the sooner these are re-planted the better, as they do not keep so well out of the ground as many others.

This Lily is considered as an emblem of purity and elegance; and

"The lady lily, looking gently down,"

is scarcely less a favourite with the poets than the rose itself:

"The lily, of all children of the spring
The palest—fairest too where fair ones are."
BARRY CORNWALL'S FLOOD OF THESSALY.

"Thus passeth yere by yere, and day by day,
Till it felle ones in a morwe of May,
That Emelie, that fayrer was to seene
Than is the lilie upon his stalke greene,
And fresher than the May with floures newe,
For with the rose color strof hire hewe;
(I n'ot which was the finer of them two)
Er it was day, as she was wont to do,
She was arisen and all redy dight;
For May will have no slogardie a-night."
CHAUCER.

————"In virgin beauty blows
The tender lily languishingly sweet."
ARMSTRONG.

"Hevinlie lyllyis with lokkerand toppis quhyte,
Opynnit and schew thare istis redemyte."
GAWIN DOUGLAS.

"Queen of the field, in milk-white mantle drest,
The lovely lily waved her curling crest."
MODERNISED BY FAWKES.

Catesby's Lily was named in honour of Mr. Catesby, who first found it in South Carolina. It is one of the smallest of the lilies cultivated in this country; the whole plant, when in bloom, being little more than a foot high. The flower is variously shaded with red, orange, and lemon colours, and has no scent. It blows in July and August. This lily does not produce offsets very fast. It must be carefully sheltered from frost, and be kept moderately moist.

The Orange-Lily has a large and brilliant flower, of a glowing flame-colour, figured and dotted with black and fiery red. There are several varieties of this species: one of which, called the Bulb-bearing Fiery Lily, puts out bulbs from the axils of the stalks; which, when the stalks decay, being taken off, and planted, will produce new plants.

The Orange-Lily will thrive in any soil or situation, and is readily increased by offsets. The bulbs should be removed every second year, and planted again before Christmas. It may stand abroad, and should be kept moderately moist.

Of the Martagon Lilies there are several species, and many varieties of each. These are not calculated for pots, but cannot be passed over without notice, since it is one of these Lilies called the Chalcedonian, or Scarlet Martagon, which has been determined to be the poetical hyacinth. (See Hyacinth.) The Red, and Yellow Martagons are commonly known by the name of Turk's-cap Lilies.

LILY OF THE VALLEY.

CONVALLARIA.

SMILACEÆ. HEXANDRIA MONOGYNIA.

These flowers are so named from growing in valleys.

Of the Lily of the Valley, called also Lily Convally, and May Lily, and in some country villages, Ladder to Heaven;—in French, *le muguet; lis des vallées; muguet de Mai*: in the village dialect, *gros mouguet*: in Italian, *il mughetto; giglio convallio* [lily convally]; *giglio delle convalli*—there are three species: the Sweet-scented, the Grass-leaved, and the Spiked. The first is a native of Britain and many other parts of Europe. It flowers in May: whence it has been named by some the May Lily. Gerarde calls it Convall Lily, and says that in some places it is called Liriconfancie. It is also called May-blossom.

"The Lily of the Valley," says Mr. Martyn, "claims our notice both as an ornamental and a medicinal plant. As an ornamental one, few are held in higher estimation: indeed, few flowers can boast such delicacy, with so much fragrance. When dried they have a narcotic scent, and, reduced to powder, excite sneezing. A beautiful and desirable green colour may be prepared from the leaves with lime." The distilled water is used in perfumery.

There are several varieties of this species: one with red flowers, one with double red, and one with double white blossoms. There is also a variety much larger than the common sort, and beautifully variegated with purple. It was brought from the Royal Garden at Paris, and flowered several years in the Chelsea Garden: but the roots do not increase so much as the other varieties.

The Lily of the Valley requires a loose sandy soil and a shady situation. It is increased by parting the roots in autumn, which should be done about once in three years. They may be gently watered every evening in dry summer weather. When the roots of this plant are confined in a pot it may also be increased by its red berry; but in the woods, where the roots are allowed to spread, it seldom produces the berry*.

The other species of the Lily of the Valley are natives of Japan.

> "No flower amid the garden fairer grows
> Than the sweet lily of the lowly vale,
> The queen of flowers."

> ————"And valley-lilies whiter still
> Than Leda's love."
>
> KEATS'S ENDYMION, p. 10.

Of the Solomon's-seal—called in French *le sceau de Salomon; le signet de Salomon; l'herbe de la rupture* [rupture-wort]; *le genouillet:* Italian, *il ginocchietto; sigillo di Salomone*—there are seven species, and varieties of each: the Narrow-leaved, the Single-flowered, the Broad-leaved, the Many-flowered, the Cluster-flowered, the Star-flowered, and the Least Solomon's-seal, or One-blade.

"The root of the Single-flowered species," says Mr. Martyn, "is twisted and full of knots. On a transverse section of it, characters appear that give it the resemblance of a seal: whence its name of Solomon's-seal." It is also called White-root.

The roots of this and the Broad-leaved kind have, in times of scarcity, been made into bread; and the young shoots of the latter species are eaten by the Turks as we eat asparagus. All the species are elegant plants. They

* See Rousseau's Letters on Botany.

are hardy; and, in a light soil and a shady situation, increase very fast by the roots. The best time to transplant them, and to part the roots, is in autumn, soon after the stalks decay. They should not be removed oftener than every third year; but should have fresh earth, as deep as it can be changed without disturbing the roots, every spring. The earth should be kept moderately moist.

There is something delightfully fresh and cool in the appearance of these Lilies; of which the flowers are so pleasantly shaded by their large light-green leaves, that one wishes one's-self a fairy to lie in them, like Ariel in the bell of the cowslip:

> "Where the bee sucks, there lurk I,
> In a cowslip's bell I lie."

It is to these Mr. Hunt alludes in one of his poems, where he seems revelling to his heart's delight among all the sweets of spring:

> "Lilacs then, and daffodillies,
> And the nice-leaved lesser lilies,
> Shading, like detected light,
> Their little green-tipt lamps of white."

LUPINE.

LUPINUS.

LEGUMINOSÆ. DIADELPHIA DECANDRIA.

The name of this plant is derived from *lupus*, a wolf, and is given it on account of its exhausting qualities.—*French*, le lupin.—*Italian*, lupino.

THE Lupines are, with one exception, annual plants; and that one strikes so deep a root, that it cannot be grown in a pot. The others may be raised without any difficulty, and are very pretty when in flower; indeed, their leaves are by no means destitute of beauty, growing in a kind of starry form, and, in most of the species, being of a downy velvet softness. The flowers are blue, white, rose-coloured, pale or deep yellow.

The Blue Lupines have usually more flowers; but the Common Yellow Lupine is often preferred for its sweet scent. The flowers of this Lupine are of very short duration, especially if the season be warm; therefore, to have a succession of them, the seed should be sown at several times; for they will continue to flower until checked by frost; and those which blow in the autumn will last longer than the earlier ones.

Lupines may be sown from the beginning of February to the end of June: they may be sown six or seven in a pot of as many inches diameter. Towards the end of June they will begin to flower. It is safer to keep such as are sown in February, in the house, until the frosts are securely over; but this precaution will not be necessary if the season be mild. They should be watered three times a week in the spring; but, as the weather becomes warmer, they may, when there is no rain, be watered every evening. About sun-set the leaves will droop as if dying, in the same manner as those of the balsam. This must not be mistaken for a want of water, as with the hydrangea: they will again display their starry foliage in the morning sun.

"Virgil calls Lupines *tristes lupini*," says Mr. Martyn, "because their bitterness contracts the muscles, and gives a sorrowful appearance to the countenance." One might rather have suspected it to have been from the drooping of the leaves, since the poets have always taken advantage of these peculiarities in flowers, to make them expressive of a sentiment, imputing a conscious timidity to the violet; and even the cowslip has, from its gentle drooping, been called in by Milton as a mourner for Lycidas. The instances of this use of flowers are without number; but there are not many more interesting than the following:

> "Bowing adorers of the gale,
> Ye cowslips, delicately pale,
> Upraise your loaded stems;
> Unfold your cups in splendour, speak!
> Who deck'd you with that ruddy streak,
> And gilt your golden gems?"
>
> CLARE.

Notwithstanding their bitterness, Lupines were formerly used as a food in some parts of Italy; particularly the White Lupine, which is a native of the Levant, and was then known also by the German name, Fig-bean. It was also called the Tame Lupine. In Tuscany and the South of France, where there is a want of manure, the Lupine is ploughed into the land, and is thought to improve it: it is used particularly for land intended for vineyards.

The Lupines are from Africa, America, and many parts of Europe. The Great Blue Lupine has been supposed to be a native of India. Parkinson says it is from Karamania; Linnæus brings it from Arabia: but, whether it comes from any, or from all of these countries, it is very hardy, and sustains the changes of our climate manfully.

LYCHNIS.

CARYOPHYLLEÆ. DECANDRIA PENTAGYNIA.

The origin of this name is not known. Some say it signifies a torch, and that the plant was so named from its flame-coloured flowers: others derive it from *lucerna*, a lamp, and suppose it to have been given from the lamp-like shape of the capsule.

THE Scarlet Lychnis, which is the handsomest of these plants, does not thrive well in a pot, because the roots naturally spread to a great distance, and do not like confinement. It may, however, be grown in a very large pot, and often is.

The double-flowered variety of the Common Meadow Lychnis is often cultivated for the beauty of the flowers. The single flowers are very common in our hedges, which are sometimes absolutely illuminated by them, in May, June, and July. They are generally known by the name of Rose-campion, but have a variety of other names; as Meadow-pink, Wild-william, Crow-flower, Cuckoo-flower, Ragged-robin, March-gilliflowers, Wild-campion, Meadow-campion, Gardener's-delight, or Gardener's-eye. This species may be increased by slipping the roots in autumn. They are of a bright rose-colour; sometimes white.

The Scarlet Lychnis has also many names; as, indeed, have most of the species. It is called by the old writers Flower of Constantinople; Flower of Bristow; Campion of Constantinople; and Nonesuch. The French call it *croix de Jerusalem; croix de Maltha; fleur de Constantinople:* the Italians, *croce di Cavalieri:* the Spaniards, *cruces de Jerusalem:* and the Portuguese, *cruz de Malta.*

The Wild Lychnis, also called Wild Campion, of which the double-flowered varieties, both red and white, are known in gardens by the name of Bachelor's-buttons, are very ornamental, continue long in flower, and blow at the same time with the Meadow Lychnis. This is increased in the same manner: the roots must be removed and parted every year. This kind, also, is sometimes called Rose-campion; but the true Rose-campion is a species of Agrostemma.

They may be kept moderately moist.

The Scarlet Lychnis is increased by cuttings, which, however, are very uncertain, and frequently fail. The cuttings should be taken from the young side-shoots, without flowers. They should have three or four joints, and be inserted to a depth half-way between the second and third. A hand-glass will facilitate their rooting.

MALLOW.

MALVA.

MALVACEÆ. MONADELPHIA POLYANDRIA.

Gerarde supposes the Latin name of this genus to be derived from the Hebrew, in which tongue it is called Malluach, from its saltness (*Melach*, salt) because the Mallow grows in salt places, among rubbish, &c., where saltpetre abounds. "I am persuaded," says he, "that the Latin word Malva comes from the Chaldee name Malluach, the ch being left out for the good sound's sake; so that in the Malua we should pronounce the *u* as a vowel, Malua, which comes near to the English word Mallow."—*French*, mauve.—*Italian*, malva.

THE Whorl-flowered Mallow is a native of China: it has pale red flowers, blowing in June and July.

The Syrian or Curled Mallow has white flowers, veined with red or purple, and is in flower from June to August. These two kinds are annual.

The Vervain and the Musk-mallow are natives of many parts of Europe, and the latter is sometimes found wild in this country.

They may be sown about the end of March, several together; and when they are three or four inches high, they may be removed into separate pots, which should be nine or ten inches wide. They may also be sown in the autumn, for they will bear the cold very well, if not too much watered; and will flower earlier, and even stronger, than those sown in the spring. In dry summer weather they may have a little water every evening, or second evening, according to the heat of the sun, the plant's exposure to it, &c.: but in cold weather, once a week, or twice in ten days, will be sufficient.

A species of Mallow was used among the Romans as an esculent vegetable. Horace mentions it as one of his ordinary dishes. We are informed that a tree of the Mallow kind furnishes food to the Egyptians, and the Chinese also use Mallows in their food. A kind of paste, called by the French name of *pâte de mauve*, was prepared from the root, which is thought to be efficacious in allaying the irritation produced by violent coughing; but at present the Mallow is omitted, that the composition may have a fine white colour; it is therefore now made only of the finest white gum-arabic, the white of eggs, sugar, and orange-flower water.

The Mallow was formerly planted, with some other flowers, the Asphodel in particular, around the graves of departed friends*. It was probably this circumstance which led to the following reflections, in the epitaph on Bion, by Moschus:

> "Raise, raise the dirge, Muses of Sicily.
> Alas! when mallows in the garden die,
> Green parsley, or the crisp luxuriant dill,
> They live again, and flower another year;
> But we, how great soe'er, or strong, or wise,
> When once we die, sleep in the senseless earth
> A long, an endless, unawakeable sleep."
>
> HUNT'S FOLIAGE.

* See Asphodel.

MARSH-MARYGOLD.

CALTHA PALUSTRIS.

RANUNCULACEÆ. POLYANDRIA POLYGYNIA.

The name Caltha signifies in Greek a basket, and refers to the appearance of the flower when not fully expanded.—*French*, le populage; le souci d'eau [water marygold]; le souci de marais; souchet d'eau. *Italian*, calta palustre; sposa del sole [spouse of the sun]; populaggine.

The Marsh-marygold makes a brilliant appearance in the meadows in March and April, and sometimes even as early as February. The flower-buds, gathered before they expand, are said to be a good substitute for capers; and their juice, boiled with alum, stains paper yellow. On May-day the country people strew these flowers before their doors, and twine them in their garlands. In Lapland it is the first flower that announces the approach of spring, although it does not there appear till the end of May. The double variety is preserved in flower gardens for its beauty, which lasts longer than in the single flowers, although blowing later. It blooms throughout May and June. This plant is increased by parting the roots in autumn: it likes the shade, and must be allowed more water than other Marygolds; for its natural place of growth is—

——————— "Not the sunny plain,
But where the grass is green with shady trees,
And brooks stand ready for the kine to quaff."

AMARYNTHUS.

MARVEL OF PERU.

MIRABILIS.

NICTAGINEÆ. PENTANDRIA MONOGYNIA.

So named from the wonderful diversity of colours in the flowers. "Every thing from the New Continent," says Mr. Martyn, "was at first esteemed wonderful."—*French,* belle de nuit [beauty of the night]; admirable de Perou; merveille de Perou; jalap faux [false jalap].—*Italian,* fior di notte [night flower]; maraviglia del Peru; sciarappa. *Spanish,* Don Diego de noche.

The Marvel of Peru, though first brought to us from that place, is also a native of Africa, China, the East and West Indies. The colours vary from white to red, purple, yellow, red and yellow, purple and yellow, purple and white, &c.; sometimes all these colours are seen in the same plant; which, being very full of blossom, has a handsome appearance. It flowers in July, August, September, and, in mild seasons, October. In warm weather the flowers do not open till the evening; but when the weather is cool, or the sun obscured, they will open in the daytime; its hours of rest being exactly the reverse of—

"*Those* flowers that turn to meet the sun-light clear,
And those which slumber when the night is near."

H. Smith.

The Japanese ladies prepare, from the meal of the seeds, a white paint for their faces.

The Forked Marvel of Peru is a native of Mexico; it is very similar to the former species, but the flowers are smaller, and do not vary in their colour, which is a red-purple, or rather a purple-red. It is very common in all the islands of the West Indies, where it is called Four-o'clock-flower, from the flowers opening at that time of the day.

The Sweet-scented Marvel of Peru has white flowers; which, as in the other two kinds, close during the day, and expand when the sun declines, like the Indian Night-flower, which its countrymen have named Sephalica, because they believe the bees sleep upon its blossoms.

The Sweet-scented species is not, however, agreeable to every one, since its scent is of musk, which many persons dislike. It is a native of Mexico, and flowers from June to September.

The Clammy Marvel of Peru has violet-coloured flowers, which blow from August to December.

These plants should be raised in a hot-bed; but if the roots are taken up when they have done flowering, laid in dry sand all the winter, secure from frost, and planted again in the spring, they will flower very well. Towards the end of May they may be set abroad. The first and third kinds are the most hardy. The earth must be but just kept moist.

If placed in a warm situation, they may be raised without a hot-bed, but will not flower till late in the season. The seeds should be sown in April, separately, in eight inch pots: when the roots are planted the second year, the pot should be rather larger.

Rousseau, in speaking of the name of this plant, says— "Upon the first discovery of the New World, as America was boastingly called, every thing found there was represented as wonderful. Strange stories were related of the plants and animals they met with, and those which were sent to Europe had pompous names given them. One of these is the Marvel of Peru, the only wonder of which is, the variety of colors in the flower."—ROUSSEAU'S LETTERS ON BOTANY.

MARYGOLD.

CALENDULA.

CORYMBIFERÆ. SYNGENESIA POLYGAMIA NECESSARIA.

The derivation of *Calendula* is uncertain: some say it is from the Calends. In English the old name for these flowers is Golds, or Rudds. Golds, or Gouldes, is a name given by the country people to a variety of yellow flowers; and the name of the Virgin Mary has been added to many plants which were anciently, for their beauty, named after Venus, of which the Marygold is one: Costmary, the Virgin Mary's Costus, is another. The French name it, souci du jardin [garden marygold]; in Provence they call it gauche fer [left-hand iron]; perhaps from its round form, like a shield which is borne on the left arm, in contradistinction to the sword, used in the right. The Italians call it calendula ortense, fiorrancio, a corruption of fiore arancio (orange flower) and fiore d'ogni mese, or flower of every month; which latter name gives countenance to the derivation of Calendula from the Calends.

THE Field Marygold is a native of most parts of Europe, and differs but little from the Garden Marygold, except in being altogether smaller.

The Garden Marygold grows naturally in the vineyards of France, the cornfields of Italy, and the orchards, fields, and gardens of Silesia. It was esteemed for its dazzling splendour long before its uses were discovered: it is a common ingredient in soups; and is said, as the old authors express it, "greatly to comfort the heart and the spirits." It has also been recommended as a medicine, but has not obtained much reputation in this way. Formerly it was considered as a wholesome ingredient in salads, but there is an acrimony in the whole plant which has even caused it to be commended as a destroyer of warts. Infused in vinegar, the Marygold is supposed to prevent infection, even that of the plague itself; and, so infused, both the leaves and flowers are found a powerful sudorific. It is, however, very probable that the efficacy of the infusion, in cases of infection, is more in the vinegar than in the flower infused in it. It has been asserted that the sting of a wasp, or a bee, is effectually cured by rubbing the part affected with a Marygold-flower.

Linnæus has observed, that the Marygold is usually open from nine in the morning to three in the afternoon. The circumstance attracted early notice, and on this account the plant has been termed *solisequa* (Sun-follower); and *solis sponsa*, Spouse of the Sun.

There is an allusion to this daily closing of the Marygold in the poems of Chatterton:

> "The mary-budde that shutteth with the light."

Another in the Pastorals of W. Browne:

> "But, maiden, see the day is waxen olde,
> And gins to shut in with the marygold."

And a most beautiful one in Shakspeare's Winter Tale:

> "The marygold, that goes to bed with the sun,
> And with him rises weeping."

And again in Cymbeline:

> "Hark! hark! the lark at heaven's gate sings
> And Phœbus 'gins arise,
> His steeds to water at those springs
> On chaliced flowers that lies.
> And winking marybuds begin
> To ope their golden eyes
> With every thing that pretty bin,
> My lady sweet arise,
> Arise, arise."

There are many varieties of the Garden Marygold; one of which, the Proliferous, called by Gerarde the Fruitful Marygold, is, as he says, "called by the vulgar sort of women, Jack-an-apes-on-horseback."

Although this Marygold is generally yellow, there is a

variety with purple flowers. The Cape Marygolds, specifically so called, as well as some others, natives of the Cape, have a deep purple centre or disk; and the florets around it, which are called the rays of the flower, are of a violet colour without, and a pure white within.

These kinds, like our common Garden Marygold, open when the sun shines, and close in the evening, and in cloudy weather. Two of these, the Grass-leaved, and the Shrubby, are perennial plants: the others are annual.

The Garden Marygold, and the Great, the Little, and the Naked-stalked Cape Marygolds, may be sown in April or in March; the first singly; the others, four of them, or five, in a pot ten inches wide. If they all come up, the two most promising should be preserved, and the rest rooted out; they will not bear transplanting. The Grass-leaved kind is best raised by a gardener; and should be housed, but not kept too warm, in the winter. The Shrubby Marygold is increased by cuttings planted in any of the summer months, and shaded from the sun until they have taken firm root, which will be in five or six weeks. In winter, this must be treated as the last.

The Marygolds must not be suffered to remain dry, but must have but little water at a time. Most of them flower from June till August; but the Garden Marygold continues in bloom till stopped by the frost.

> "Open afresh your round of starry folds
> Ye ardent marigolds!
> Dry up the moisture of your golden lids,
> For great Apollo bids
> That in these days your praises shall be sung
> On many harps, which he has lately strung;
> And when again your dewiness he kisses,
> Tell him, I have you in my world of blisses:
> So haply when I rove in some far vale,
> His mighty voice may come upon the gale."
>
> Keats.

It has been observed that these flowers were formerly called Golds, a name by which Chaucer repeatedly mentions them: we are told, in the glossary, that Gold means a Sun-flower, but it has been remarked that this title also was formerly bestowed upon the Marygold: and the following passage is an additional argument for supposing Chaucer to have intended this flower rather than the enormous Sun-flower, now so called:

> "Eke eche at other threwe the flouris bright,
> The prymerose, the violete, and the gold."
>
> COURT OF LOVE.

He also bestows a garland of them upon Jealousy, yellow being the colour emblematical of that passion:

> ———————"and Jalousie,
> That wered of yelwe goldes a gerlond,
> And had a cuckowe sitting in her hand."
>
> THE KNIGHT'S TALE.

MAURANDIA SEMPERFLORENS.

SCROPHULARIEÆ. DIDYNAMIA ANGIOSPERMIA.

THIS shrub is a native of Mexico, and requires winter shelter. It should be housed at Michaelmas, or, if the season be cold, somewhat earlier; and should remain within, till the end of May, or the beginning of June. In April and in October the earth should be removed as deep as can be done without disturbing the roots, and fresh earth substituted. If it requires new potting, it must be carefully removed with the ball of earth about the roots: all the matted, decayed, or mouldy roots on the outside, should be pared away; and, when fresh planted, it should be gently watered, and placed in the shade. April is the best time for transplanting this shrub, that it may have time to fix its roots before the time of its removal into the open air. In winter it will not require water oftener than once in five days or a week; and then, unless in very mild weather, at the roots only. In dry summer weather it may be watered every evening, or second evening, according to its situation and the heat of the sun.

MESEMBRYANTHEMUM.

FICOIDEÆ. ICOSANDRIA PENTAGYNIA.

The name of this genus is derived from three Greek words, and signifies a flower with the embryo in the middle: it was originally named Mesembrianthemum, or Noon-flower, because most of the species close in the absence of the sun, and disclose themselves in broad sunshine. The familiar name is Fig-marygold.—*French,* ficoïde; fleur du midi [noon-flower].—*Italian,* ficoide.

THIS genus is very large: it will be necessary only to select a few of the more desirable or general kinds.

One of the most popular kinds of the Mesembryanthemums is the M. Crystallinum, Ice-plant, or Diamond-fig Marygold, of which the leaves, stems, and buds are apparently covered with ice.—French, *la glaciale:* Italian, *erba crystallina.*—It comes from the neighbourhood of Athens. This plant must be raised in a hot-bed; but, if not intended for seeding, may be placed abroad in May, and will preserve its beauty till late in the autumn. It is chiefly for the ice-like surface that it is admired: the flowers are trivial. It must be very cautiously watered, and only often enough to preserve the earth from becoming an absolute dust.

The M. Frequentiflora is, as its specific name implies, a frequent flowerer, but it must not be kept too warm, or it will not satisfy this expectation; it must merely be sheltered from frost and wintry winds.

The M. Variabile flowers in June, July, and August. It frequently changes its colour; on first opening, it is orange or saffron-coloured, then yellow, which fades almost to a white, with a tinge of red, and a red midrib, and at last a fleshy white, and rubicund on the outside. The M. Versicolor is rubicund in the morning, pale silvery in the middle of the day, and rubicund again in the evening.

M. Polyanthon makes a brilliant show in August. M. Emarginatum flowers from June to August. M. Speciosum is one of the most beautiful of them all, and flowers freely. M. Floribundum flowers from May to October in a most exuberant manner. M. Striatum blossoms also from May to October very liberally; and so great is the demand for this species, that a nursery-florist, near Hammersmith, has been said to sell 7000 pots of it annually.

M. Bracteatum flowers from August to January; M. Stellatum from August to November; M. Inclaudens flowers in July: the flowers are always expanded, and very handsome.

M. Decumbens, from May to October; M. Glomeratum from June to July; and M. Falciforme from July to August; are very full of blossom. The flowers of M. Caulescens smell like new hay: this flowers in May.

These plants should be housed in the winter, and carefully sheltered from the frost; but fresh air must always be allowed them in mild weather. They are very succulent, and must have but just water enough to keep them alive: the less moisture is given them, the better they will bear the winter cold.

M. Lacerum is more tender than most of the genus, and must therefore be sooner housed, and set further from the windows when open, in the winter. The flowers of this kind are of a pale rose-colour, large and showy, and do not close at night. It flowers in June.

M. Tenuifolium has pale red flowers; which, in the sunshine, appear sprinkled with gold-dust: they blow very abundantly in June. Most of the species have purple or yellow flowers.

It may be observed, as a general rule, to water the shrubby kinds twice, the more succulent, once a week, during the summer; but, towards the end of autumn, it must be given less frequently: once a week to the shrubby kinds, when the weather is not frosty; and in severe weather, the succulent should have no more water than just to prevent their leaves from shrivelling.

They are best raised by a gardener. The pots should be frequently examined at the bottom, to see if the roots run through, in which case they must be cut off. Those perennial kinds which grow pretty fast should be shifted once or twice in the course of the summer, to pare off their roots, and, if necessary, remove them into larger pots; but they should be always kept in as small pots as possible, particularly those of the more succulent kind.

They should generally be housed in September, and placed abroad in May, in a sheltered, warm, sunny situation. In very wet weather, the most succulent kinds should be screened from it.

This is a handsome and admired genus, and comprehends a great variety. They are chiefly natives of the Cape. Few green-houses, however small, are without the Ice-plant; which is also, from its glittering surface, called the Diamond-plant, Diamond-ficoides, and Spangled-beau:

————————" Geranium boasts
Her crimson honours, and the spangled beau,
Ficoides, glitters bright the winter long.
All plants of every leaf, that can endure
The winter's frown, if screened from his shrewd bite,
Live there and prosper."

COWPER.

MIGNONETTE.

RESEDA ODORATA.

RESEDACEÆ. DODECANDRIA TRYGYNIA.

THIS plant is supposed to be an Egyptian, and to have been brought hither from the South of France, where it is called *reseda d'Egypte*, and *herbe d'amour* [love flower]. A French appellation, derived from the Spanish, *minoneta*, prevails here over its classical one. It is a favourite plant, very fragrant, and has well justified this affectionate name, Mignonette, or Little-darling: its sweetness wins all hearts.

"The luxury of the pleasure-garden," says Mr. Curtis, "is greatly heightened by the delightful odour which this little plant diffuses; and as it grows more readily in pots, its fragrance may be conveyed into the house. Its perfume, though not so refreshing perhaps as that of the Sweet-briar, is not apt to offend the most delicate olfactories. It flowers from May to the commencement of winter."

People have not been satisfied, however, with growing this little-darling in pots; it is more frequently seen cradled in the sunshine, in boxes the whole length of the window it is placed in.

> ——" the sashes fronted with a range
> Of orange, myrtle, or the fragrant weed,
> The Frenchman's darling."——
>
> COWPER.

The seeds may be sown in April, and will grow very well in the open air, although it will not flower so early as when raised in a hot-bed; they will, however, be much stronger. If sheltered in the winter, it will continue flowering most part of the year, but will not be so strong the second year as the first. It is generally an annual. The earth should be kept moderately moist.

MILK-WORT.

POLYGALA.

POLYGALEÆ. DIADELPHIA OCTANDRIA.

Polygala is from the Greek, and signifies much milk: the plants, when eaten by cattle, being supposed to make them yield much milk. *French,* l' herbe à lait; laitier.—*Italian,* poligala; all of a similar signification.

THE Polygala-myrtifolia, or Myrtle-leaved Milk-wort, is a shrub growing three or four feet high. It produces at the ends of the branches, red flowers, white on the outside, and of a bright purple within; and, as it is in blossom most part of the summer, is much esteemed to adorn drawing-rooms, balconies, &c.

This shrub must be housed at the approach of winter, about the middle or end of October. It must be constantly but sparingly watered; less in the winter, than when exposed to the open air in summer.

The Spear-leaved kind, which, like the former, is a native of the Cape, is a very pretty plant, flowering from May to July, and may be treated in the same manner.

MIMOSA.

LEGUMINOSÆ. POLYGAMIA MONOECIA.

Commonly called the Sensitive-plant.—*French,* herbe vive [live herb]; herbe sensible; herbe sensitive; acacia.—*Italian,* sensitiva.

SOME few species of the Mimosa may be preserved in a warm inhabited room; but they are mostly kept in a stove, and few of them will bear the open air even in summer. Like human beings, they are more sensitive in proportion to the tenderness of their nursing: like them, by living hardily, they may be fitted to bear the common chances of life. In the plant, this nervous sensibility is encouraged for its singularity; it is pity there should not be the same reason for encouraging it in the human species.

If the roots shoot through the pot at the bottom, the plant should be turned out, the roots be pared close, and then replaced in the same pot, or a larger, if necessary; but they do not thrive so well in large pots. Great caution must be observed in watering them; they must have little water at a time, but must not be suffered to remain quite dry.

Many persons have endeavoured to ascertain the cause of the sensibility of these plants, but it has never yet been clearly explained. The degree varies in the different kinds: some will only contract their leaves on being touched; others will bend and recede, as it were courteously to acknowledge your approach; as that which is termed the Humble-plant.

————————" that courteous tree
Which bows to all who seek its canopy."
T. Moore.

" Looke as the Feeling-plant, which learned swaines
Relate to growe on the East Indian plaines,
Shrinkes up his dainty leaves if any sand
You throw thereon, or touch it with your hand."
W. Browne

The most irritable part of the plant is in the foot-stalk, between the stem and the leaflet. During the night they remain in the same state as when touched in the daytime; yet, if touched then, will fold their leaves still closer.

When any of the upper leaves are touched, if in falling they touch those below them, these also will contract and fall; so that by touching one another, they will continue to fall for some time. Mimosas are very common in the woods of Brazil: of one of the species, of which the wood is very light, the Indians make their canoes*.

A species of the Mimosa, called the Egyptian Mimosa, or Egyptian Acacia, produces the gum-arabic. It is a native of Egypt and Arabia, and, in its own countries, grows to a considerable size. This tree was called by Theophrastus, an Acanthus; and is spoken of by Virgil under the same name, in the second Georgic:

"Quid tibi odorato referam sudantia ligno
Balsamaque, et baccas semper frondentis acanthi?"

GEORGIC 2.

"Why should I mention the balsam which sweats out of the fragrant wood, and the berries of the ever-green acanthus?"

MARTYN'S TRANSLATION.

In this passage he is supposed to refer to the Egyptian Mimosa.

In the fourth Eclogue, where anticipating a golden age, he speaks of the Acanthus as one of the plants which the earth is to bear without trouble:

"At tibi prima, puer, nullo munuscula cultu,
Errantes hederas passim cum baccare tellus,
Mistaque ridenti colocasia fundet acantho."

"Meanwhile the earth, sweet boy, as her first offerings, shall pour thee forth every where, without culture, creeping ivy with ladies-glove, and Egyptian beans with smiling acanthus intermixed."

DAVIDSON'S TRANSLATION.

In the third, where he wreathes the Acanthus round the handles of Alcimedon's cups, and in the fourth Georgic, where he places it in the Corycian's garden, he alludes to the herb Acanthus, commonly called, from its roughness,

* See Prince Maximilian's Travels in Brazil. This author speaks of these Mimosas as growing to a size even colossal, and diffusing a delicious perfume; which, together with that of the magnificent creeping plants generally interwoven with their branches, attracts a great number of butterflies and humming-birds, which hover about them like bees.

Branca-ursi, or Bear's-breech. This Dryden has translated Bear's-foot, which is a very different plant; a species of Helleborus.

As this last passage applies, in a general as well as particular manner, to the work now before us, we will quote some lines from Dryden's translation. It immortalizes an old acquaintance of the poet's, who was a gardener:

"Now did I not so near my labours end,
Strike sail, and hastening to the harbour tend,
My song to flowery gardens might extend.
To teach the vegetable arts, to sing
The Pæstan roses, and their double spring;
How succory drinks the running streams, and how
Green beds of parsley near the river grow;
How cucumbers along the surface creep,
With crooked bodies, and with bellies deep,
The late narcissus, and the winding trail
Of bear's-foot, myrtles green, and ivy pale:
For where with stately towers Tarentum stands,
And deep Galæsus soaks the yellow sands,
I chanced an old Corycian swain to know,
Lord of few acres and those barren too,
Unfit for sheep or vines, and more unfit to sow:
Yet, labouring well his little spot of ground,
Some scattering pot-herbs here and there he found,
Which cultivated with his daily care,
And bruised with vervain, were his frugal fare.
Sometimes white lilies did their leaves afford,
With wholesome poppy-flowers to mend his homely board:
For, late returning home, he supped at ease:
The little of his own, because his own, did please.
To quit his care, he gathered first of all
In spring the roses, apples in the fall:
And when cold winter split the rocks in twain,
And ice the running rivers did restrain,
He stripped the bear's-foot of its leafy growth,
And calling western winds, accused the spring of sloth.
He therefore first among the swains was found
To reap the product of his laboured ground,
And squeeze the combs with golden liquor crowned.
His limes were first in flower; his lofty pines
With friendly shade secured his tender vines,

For every bloom his trees in spring afford,
An autumn apple was by tale restored.
He knew to rank his elms in even rows,
For fruit the grafted pear-tree to dispose,
And tame to plums the sourness of the sloes.
With spreading planes he made a cool retreat,
To shade good fellows from the summer's heat;
But, straitened in my space, I must forsake
This task, for others afterwards to take."

The Acanthus was one of the most favourite ornaments of the Greeks; and, as is well known, makes the principal figure in the capital of the Corinthian column; the idea of which is said to have been suggested by the accidental sight of a basket overgrown by Acanthus, with a tile on it.

Martyn's notes to Virgil's Georgics contain some very interesting remarks on both the kinds of Acanthus mentioned by that poet; and he quotes a passage from Vitruvius, on the origin of the use of the Acanthus in architecture: "This famous author tells us, that a basket covered with a tile having been accidentally placed on the ground, over a root of acanthus, the stalks and leaves burst forth in the spring, and spreading themselves on the outside of the basket, were bent back again at the top by the corners of the tile. Callimachus, a famous architect, happening to pass by, was delighted with the novelty and beauty of this appearance; and, being to make some pillars at Corinth, imitated the form of this basket, surrounded with acanthus, in the capitals. It is certain there cannot be a more lively image of the capital of a Corinthian pillar than a basket covered with a tile, and surrounded by leaves of brank-ursine, bending outward at the top."—Others say that the acanthus of the architects is a different species, though of the same genus with the brank-ursine.

Virgil again mentions an Acanthus as forming the pattern upon a mantle which had belonged to Helen; and Theocritus, as a relievo upon a pastoral prize-cup:

> "And all about the cup a crust was raised
> Of soft acanthus."

But these Acanthuses either will not bear our climate, or they strike their roots too deep for potting; so that, however interesting they are rendered by classical association, it is in books we must enjoy them, if we would enjoy them at home.

MINT.

MENTHA.

LABIATÆ. DIDYNAMIA GYMNOSPERMIA.

Supposed to be named from Mentha, a daughter of Cocytus, who was changed into this herb by Proserpina in a fit of jealousy. (See Anemone.)

Mint may seem to belong rather to a kitchen or a physic, than to a flower garden; but besides its medicinal and culinary uses, Spear-mint is esteemed by many persons for the scent; and for this scent, and the quickness of its growth, it is often grown in pots, although as a flower it is of no value. This mint is called in French, *menthe verte* [green mint]; *menthe d Angleterre* [English mint]; *menthe Romaine* [Roman mint]; *menthe de Nôtre Dame* [Our Lady's mint]: in Italian, *erba Santa Maria* [the Holy Mary's herb]. The flower, such as it is, is of a dark purple; the leaves are handsome, and, when they grow luxuriantly, have a cool and refreshing appearance. Preparations from several of the species are used in medicine; as Pepper-mint, Spear-mint, Pennyroyal, &c.

An infusion of Spear-mint is used as a substitute for tea; the young leaves are eaten in salads, and some eat them in the same manner as the leaves of sage, with bread and butter.

All the Mints are easily increased by parting the roots in the spring, or by planting cuttings in any of the summer months; keeping the earth very moist until they have taken root.

Mint to be used as tea, should be cut when just beginning to flower, and dried in the shade.

It is said that Corn-mint—in French, *le pouliot thym* [thyme pennyroyal]—prevents the coagulation of milk; and "when cows have eaten it," says Withering, "as they will do largely at the end of summer, when pastures are bare, their milk can hardly be made to yield cheese; a circumstance which puzzles the dairy-maids."

MONK'S-HOOD.

ACONITUM.

RANUNCULACEÆ. POLYANDRIA TRIGYNIA.

Called also Wolf's-bane; Aconite.—*French,* l'aconit: in some places, toutchoz.—*Italian,* aconito.

The Monk's-hoods are hardy perennials, very handsome; their flowers growing in spikes or rods, which, in some of the species, are nearly two feet long. They may be increased by parting the roots, every piece of which will grow. This should be done soon after they have done flowering; and the stalks should be cut down at the same time. They like shade, and moisture. Most of them have blue flowers, but there are also white and yellow.

The ancients, who were not acquainted with chemical poisons, regarded the Aconite as the most violent of all; and fabled it to be the invention of Hecate, and sprung from the foam of Cerberus:

"And now arrives unknown Ægeus' seed,
Who, great in name, had two-sea'd Isthmos freed;

Whose undeserved ruin Phasias * sought
By mortal aconite, from Scythia brought:
This from th' Echidnean dog dire essence draws.
There is a blind steep cave, with foggy jaws,
Through which the bold Tyrinthian hero† strain'd,
Dragg'd Cerberus, with adamant enchain'd;
Who backward hung, and scowling, look'd askew
On glorious day, with anger rabid grew;
Thrice howls, thrice barks at once with his three heads,
And on the grass his foamy poison sheds.
This sprung: attracting from the fruitful soil
Dire nourishment, and power of deathful spoil.
The rural swains, because it takes delight
In barren rocks, surnamed it aconite."

SANDYS'S OVID.

The real virulence of the Aconite has been proved by fatal experience. Some persons, only by smelling the full-blown flower, are said to have been seized with swooning fits, and to have lost their sight for two or three days. The root is the most powerful part of the plant; and a criminal has been put to death by being made to swallow one drachm of it. Dodonæus mentions an instance, recent in his time, of five persons at Antwerp who ate of the root by mistake, and all died. Instances have also been recorded of persons who have died from eating of this in a salad, instead of celery; and some experiments upon animals have been made with it, which are too horrible to repeat.

Yet, when used with skill and caution, this plant has been found in some cases a useful medicine. Those with blue flowers are considered as the most powerful.

There is a species called the Wholesome Aconite—in French, *maclou*—which has been recommended as an antidote to the poisonous kinds. This, however, is poisonous, though not so powerfully so as the others. All the kinds have rather deep roots, which render them unfit for grow-

* Medea. † Hercules.

ing in pots. They are here mentioned rather to speak of their dangerous tendency than to recommend their increase.

MOTHERWORT.

LEONURUS.

LABIATÆ. DIDYNAMIA GYMNOSPERMIA.

The common kind is called in *French,* l'agripaume; la cordiale.—*Italian,* agripalma; cardiaca.

The Common Motherwort, and the Curled, when the seeds are once sown in the spring, require no further care than occasionally to water them, and to keep them clear of weeds. The roots last many years: they are usually of a pale red in the first, pure white in the latter kind.

The Small-flowered, Tartarian, and Siberian, require no more care than the other kinds, but to sow them oftener. They do not blow till the second year, and blow but once. Their colours are flesh-coloured, yellow, and red: blowing from June to August.

In a garden they sow their own seeds; and, when once introduced, give no further trouble.

In Japan, the Motherwort is in great estimation. It seems there was formerly, to the north of the province of Nanyo-no-rekken, a village situated near a hill covered with Motherwort flowers. At the foot of the hill was a valley, through which ran a stream of pure water, formed by the dew and rains that trickled down the sides of the hill. This water was the ordinary beverage of the villagers, who generally lived to the age of a hundred, or a hundred and thirty. To die at seventy was considered as a premature death. Thus the people have still an idea that the Motherwort has the property of prolonging life. At the court of the Dairi, the ecclesiastical sovereign of Japan, they amuse themselves with drinking zakki*, prepared from these flowers.

The Japanese have five grand festivals in the course of the year. The last festival, which takes place on the 9th of the ninth month, is called the Festival of Motherwort; and the month itself is named *Kikousouki*, or Month of Motherwort-flowers. It was formerly the custom to gather these flowers as soon as they had opened, and to mix them with boiled rice, from which they prepared the zakki used in celebrating this festival. In the houses of the common people, instead of this beverage, you find a branch of the flowers fastened with a string to a pitcher full of common zakki; which implies, that they wish one another a long life. The origin of this festival is as follows:

It is related by several authors, that a Chinese emperor, who succeeded to the throne at seven years of age, was distressed by a prediction that he would die before he attained the age of fifteen. An immortal having brought to him, from Nanyo-no-rekken, a present of some of the beautiful yellow flowers of the Motherwort, he caused zakki to be made from them, which he drank every day, and lived upwards of seventy years.

This immortal had been in his youth in the service of the emperor, under the name of Zido. Being banished for some misdemeanour, he took up his residence in the valley before-mentioned, drinking nothing but the water impregnated with these flowers, and lived to the age of three hundred years; whence he obtained the name of Sien-nin-foso.

The Japanese are, indeed, very fond of flowers in ge-

* Zakki is a kind of strong beer; the common beverage in Japan.

neral; and the houses of respectable people have always pots of flowers in the windows. They have a great esteem for plum and cherry trees, and for the beauty of their blossoms. Some dwarf trees of these kinds are cultivated in boxes behind the houses almost invariably; and persons in easy circumstances have in their apartments one or more branches, when in flower, in a porcelain vase *.

MYRTLE.

MYRTUS.

MYRTEÆ. ICOSANDRIA MONOGYNIA.

So named from Myrsine, an Athenian damsel, and favourite of Minerva, who was metamorphosed into this shrub, which is consecrated to Venus. The connexion between Minerva's favourite and Venus is not clear; but nothing can be fitter for rendering sacred to Venus than the myrtle. It is the perfection of neatness and elegance, and leaf and flower are alike worthy of each other.—*French,* le myrthe; myrte.—*Italian,* mirto.

THE Myrtle is a native of Asia, Africa, and the South of Europe; and though not very tender, is not quite hardy enough to bear our winters without some protection, except in the most southern and western parts of the island.

The beauty and fragrance of the flower are exquisite: it blossoms in July and August, but does not bear fruit in England.

The Common Broad-leaved or Roman Myrtle does not grow higher in England than eight or ten feet; but in Italy it grows much higher, and is the principal under-

* See Titsingh's Illustrations of Japan, translated from the French by F. Shoberl.

wood of some of the forests. The flowers of this are larger than those of the other species, and it is by some called the Flowering Myrtle, because it flowers more freely in this country than most others.

There are many varieties of the Common Myrtle; as the Italian, the Orange-leaved, the Bay-leaved, the Broad-leaved Dutch, the Box-leaved, the Thyme-leaved, the Double-flowering, &c. The Box-leaved has very small blossoms, which blow late in the summer.

All the varieties of this Myrtle may be increased by cuttings. The most straight and vigorous young shoots should be selected: they should be six or eight inches long, and the leaves should be stripped off two or three inches high. The part which is put into the earth should be a little twisted: they should be planted two inches apart, the earth pressed close to them, and a little water given them. They must be shaded from the noon-day sun, and be kept always moderately moist.

These cuttings should be planted in July: they are generally placed in a hot-bed, but will take root very well without that assistance, though not so quickly as where it can be allowed them. They should, however, be removed within doors, admitting fresh air. With the exception of the Orange-leaved and the Nutmeg Myrtles, which are somewhat tenderer than the rest, they may have air given them in mild weather throughout the winter: only requiring protection from frost.

During the winter, they should be gently watered twice or thrice a week, when not frosty. If any decayed leaves appear, they should always be pinched off.

The young plants should be carefully taken up in the spring, with the ball of earth adhering to their roots, and parted into separate pots; watered well, and kept in the house till they have again fixed their roots. About the middle of May they should be gradually accustomed to the open air, but placed where they may be defended from strong winds.

During the summer, Myrtles require plenty of water, especially the young plants, which, being in small pots, are sooner dry. They should be so placed as to receive the morning sun only; for if they are exposed to the meridian heat, the moisture contained in these small pots will soon be exhaled, and the growth of the plants much retarded.

In August, if the roots have made their way through the bottom of the pot, the plant must be removed into one a size larger: the cultivator paring off the decayed and matted roots, and with the hand loosening some of the earth adhering to them, that they may find an easier passage into the fresh earth. When newly planted they must always be well watered, and placed in the shade. If the branches grow in an irregular and unsightly manner, this is a good time to trim them. But the sort with double flowers should not be clipped, because the chief beauty consists in the blossoms, and the cropped branches will not produce any.

Myrtles should be first planted in very small pots, and removed into larger, as the increase of the roots may require; but large pots will not only weaken, but sometimes destroy them; so that they must not be removed into larger than they really want. The best seasons for removing them are in April and in August. About the middle, or, if the season be mild, the end of October, Myrtles should be removed into the house, both old and young; and gradually returned to their out-door station in April or May.

In Cornwall and Devonshire, where the winters are milder than in most other parts of England, Myrtles will endure the open air all the year round; and there are Myrtle hedges which have grown to a considerable height, and are very strong and healthy. Mr. Keppel Craven describes the hedges in Naples to be as commonly composed of Myrtles and orange trees, as ours are of thorn and privet. Their fragrance, when in blossom, must surpass even our own hawthorn.

The Myrtle was formerly used in medicine: it was a great favourite with the ancients; and either on account of its beauty, or because it thrives best in the neighbourhood of the sea, it was held sacred to Venus—as the olive to Minerva, the poplar to Hercules, the ivy and the vine to Bacchus, the hyacinth and the bay to Apollo, &c.

Myrtle-berries were used in cookery; and both those and the branches put into wine. Evelyn speaks of a decoction of Myrtle-berries for dying the hair black.

Myrtle was the symbol of authority for magistrates at Athens; bloodless victors were crowned with Myrtle; and hence the swords of Harmodius and Aristogiton were wreathed with Myrtle, when they set forth to free their country from hereditary monarchy. Thus when the young hero is contemning the indolent and effeminate luxury around him, he breaks out in enthusiastic admiration of the Greeks:

"It was not so, land of the generous thought,
And daring deed! thy godlike sages taught;
It was not thus, in bowers of wanton ease,
Thy Freedom nursed her sacred energies;
Oh! not beneath the enfeebling, withering glow
Of such dull luxury did those myrtles grow
With which she wreathed her sword when she would dare
Immortal deeds; but in the bracing air
Of toil, of temperance, of that high, rare,
Etherial virtue, which alone can breathe
Life, health, and lustre into Freedom's wreath."

MOORE'S LALLA ROOKH.

The Myrtle's fondness for the sea-shore is noticed by Virgil in his Georgics:

——————" nec sera comantem
Narcissum, aut flexi tacuissem vimen acanthi,
Pallentesque hederas, et amantes litora myrtos."
GEORGIC 4.

" Nor had I passed in silence the late-flowering daffodil, the stalks of the flexile acanthus, the pale ivy, or the myrtle that loves the shore."——DAVIDSON'S TRANSLATION.

And again:

" Litora myrtetis lætissima."

The same poet, in his Pastorals, alludes to the fragrance of the Myrtle-blossom:

" Et vos, ô lauri, carpam, et te, proxima myrte;
Sic positæ quoniam suaves miscetis odores."

" And you, ye laurels, I will crop; and thee, O myrtle, next *in dignity to the laurel;* for thus arranged, you mingle sweet perfumes."

So Davidson translates this passage: the words in Italics marking an interpolation, or rather a necessary explanation of the preceding adjective.

It was impossible that Spenser should omit the Myrtle in the garden of Adonis:

" Right in the middest of that paradise
There stood a stately mount, on whose round top
A gloomy grove of myrtle-trees did rise,
Whose shady boughs sharp steel did never lop,
Nor wicked beasts their tender buds did crop;
But like a garland compassed the height.
And from their fruitful sides sweet gum did drop,
That all the ground with precious dew bedight,
Threw forth most dainty odours and most sweet delight.

And in the thickest covert of that shade
There was a pleasant arbour, not by art,
But by the trees' own inclination made;

Which knitting their rank branches part to part,
With wanton ivy-twine entrail'd athwart;
And eglantine and caprifole among,
Fashion'd above within their inmost part,
That neither Phœbus' beams could through them throng,
Nor Æolus' sharp blast could work them any wrong."

FAIRY QUEEN, b. iii. c. 6.

The Myrtle and the bay are continually coupled together by the poets, like the lily and the rose. And not even the bay itself has been more sweetly sung than this beautiful shrub:

" And in the midst of all, cluster'd about
With bay and myrtle, and just gleaming out,
Lurk'd a pavilion,—a delicious sight,
Small, marble, well-proportion'd, mellowy white,
With yellow vine-leaves sprinkled, but no more,
And a young orange either side the door."

STORY OF RIMINI.

" Never look'd the bay so fit
To surmount two eyes of wit,
Nor the myrtle to be seen
Two white kerchief'd breasts between,
Nor the oak to crown a sword
For a nation's rights restored."

DESCENT OF LIBERTY.

" A sacred hedge runs round it; and a brook,
Flowing from out a little gravelly nook,
Keeps green the laurel and the myrtle trees,
And odorous cypresses."

HUNT'S FOLIAGE: *from Theocritus.*

There is another most exquisite passage about this shrub in Keats's Sleep and Poetry:

———————" a myrtle, fairer than
E'er grew in Paphos, from the bitter weeds
Lifts its sweet head into the air, and feeds
A silent space with ever-sprouting green.
All tenderest birds there find a pleasant screen,

Creep through the shade with noisy fluttering,
Nibble the little cupped flowers, and sing.
Then let us clear away the choaking thorns
From round its gentle stem; let the young fawns,
Yeaned in after-times, when we are flown,
Find a fresh sward beneath it, overgrown
With simple flowers."

"Like a myrtle tree in flower
Taken from an Asian bower,
Where with many a dewy cup
Nymphs in play had nursed it up."

HUNT: *from Catullus.*

It has been observed, that the Myrtle is consecrated to Venus. Drayton, in his Muses' Elysium, has assembled a number of emblematical wreaths:

"The garland long ago was worn,
As Time pleased to bestow it:
The laurel only to adorn
The conqueror and the poet.
The palm his due, who, uncontroll'd,
On danger looking gravely,
When fate had done the worst it could,
Who bore his fortunes bravely.
Most worthy of the oaken wreath
The ancients him esteem'd,
Who in a battle had from death
Some man of worth redeem'd.
About his temples grace they tie,
Himself that so behaved,
In some strong siege by th' enemy
A city that hath saved.
A wreath of vervain heralds wear,
Amongst our garlands named,
Being sent that dreadful news to bear,
Offensive war proclaim'd.
The sign of peace who first displays
The olive wreath possesses;
The lover with the myrtle sprays
Adorns his crisped tresses.

In love the sad forsaken wight
  The willow garland weareth,
The funeral man, befitting night
  The baleful cypress beareth.
To Pan we dedicate the pine,
  Whose slips the shepherd graceth;
Again the ivy, and the vine,
  On his swoln Bacchus placeth."

NARCISSUS.

NARCISSEÆ. HEXANDRIA MONOGYNIA.

Named from the youth Narcissus, who, as the poets tell us, was changed into this flower. Also named Daffodil. Some of the species are called Jonquils.

THE Two-flowered Narcissus, Pale Daffodil, or Primrose-peerless, is of a pale cream-colour, with a yellow cup in the centre. It grows wild in England and many other parts of Europe, and flowers in April.

Of the Common Daffodil there are many varieties: with a white flower, and yellow cup; a yellow flower, and deep golden cup; a double flower, with several cups, one within the other; Tradescant's Daffodil, "which," says Mr. Martyn, "may well be entitled the Prince or Glory of Daffodils;" the Great Nonsuch; the Great Yellow Incomparable Daffodil, which, when double, is called by gardeners, Butter-and-egg Narcissus. It is called in the Dutch catalogues, the Orange Phœnix, and is considered the handsomest of all the varieties. There are many others, which it is not necessary to specify. They mostly flower in April. This in France has many names: as, *le narcisse sauvage; le faux narcisse; campane jaune* [yellow bell]; *aiau; aioult.* In Italian, *narcisso giallo* [yellow narcissus].

The Sweet-scented Narcissus, or Great Jonquil, is a native of the South of Europe. Most of the species are fragrant; but this is the most powerful, and is often found too much so to be endured in a room.

There is a species called the Hoop-petticoat Narcissus, of which the cup is two inches long, very broad at the brim, and is said to be formed like the old bell-hoop-petticoat formerly worn by ladies in this country.

The Polyanthus Narcissus—called in France *la narcisse de Constantinople;* in Languedoc, *pissauleich:* in Italy, *tazetta*—grows naturally in the East, and in many parts of Europe. There are more varieties of this than of any other species. That which is generally called the Cyprus Narcissus, with very double flowers, the outer petals white, the inner, some white and some orange, is the most beautiful of them all, and the most esteemed for blowing in glasses in a room. Its scent is very agreeable, and less powerful than that of the Jonquil.

The White, or Poetical Narcissus,—called by the French *janette des contois*—has a snow-white flower, with a yellow cup in the centre, fringed on the border with a circle of bright purple. It is sweet-scented, a native of many parts of Europe, and flowers in May. There is a variety with double flowers.

There is a species of Narcissus which is called the Late-flowering, and does not blow till autumn. The Common Jonquil is altogether yellow, as is also the Sweet-scented; but the latter has the cup somewhat deeper coloured than the petals.

The preferable kinds are the Polyanthus Narcissus, the Jonquil, and the Poetical Narcissus; but any of them may be blown, either in glasses or pots, without difficulty, and may be readily increased by offsets.

Although it has been observed that most of these flowers blow in April and May, this only applies to such as are left in the earth to blow at their own season; but, according to their time of planting and their situation, they may be continued for many months in succession. Those planted in pots should be covered an inch over the top of the bulb; and the pot should not be less than seven inches in depth. According to the size of the bulb, one or more may be planted in each pot. They may be planted any time from September to February. Careful admissions of air in mild weather will be beneficial; and they must on no account be denied the enjoyment of daylight and sunshine, towards which they will lean with an almost animal yearning, which it were a sort of cruelty not to indulge.

Water must not be given them until the green begins to appear: they should then be gently watered once or twice a week. In a warm inhabited room they may be blown even in the midst of winter.

Such as are blown in glasses should have fresh water about once in ten days. The leaves should never be plucked off before they decay, or the root will be thereby deprived of much of its natural nourishment. When they have decayed, the bulbs should be taken up, laid in the shade to dry, cleaned, and put in a dry secure place till wanted to re-plant. The offsets should be taken off, and sorted according to their size. When planted, they may be put two or three together, until they have grown large enough for flowering.

When the plants are somewhat advanced in height they will require a stick to support them. Such plants as are kept in the open air in the spring must be defended from strong winds, which would otherwise be apt to break the stems, particularly after rains; when their cups, being filled with water, will be more heavy:

"All as a lily pressed with heavy rain,
Which fills her cups with showers up to the brinks,
The weary stalk no longer can sustain
The head, but low beneath the burden sinks."

P. Fletcher.

They will thrive best in a south-eastern exposure, where the morning sun may dry off the moisture which has lodged upon them during the night; and they will better preserve their beauty there than in the shade, or in the scorching heat of the afternoon sun.

The poetical origin of this flower, and its own beauty, have conspired to obtain for it the notice of some of the greatest poets. The story told at length in Ovid's Metamorphoses, of the transformation of Narcissus into a flower, is too well known to need, and too long to admit of, insertion.

The Naiades, lamenting the death of Narcissus, prepare a funeral pile, but his body is missing:

"Instead whereof a yellow flower was found,
With tufts of white about the button crown'd."

Sandys's Ovid.

"What first inspired a bard of old to sing
Narcissus pining o'er the untainted spring?
In some delicious ramble he had found
A little space, with boughs all woven round;
And in the midst thereof a clearer pool
Than e'er reflected in its pleasant cool
The blue sky here and there serenely peeping
Through tendril wreaths fantastically creeping.
And on the bank a lonely flower he spied,
A meek and forlorn flower with nought of pride,
Drooping its beauty o'er the watery clearness,
To woo its own sad image into nearness.
Deaf to light Zephyrus, it would not move;
But still would seem to droop, to pine, to love.
So, while the poet stood in this sweet spot,
Some fainter gleamings o'er his fancy shot;
Nor was it long ere he had told the tale
Of young Narcissus, and sad Echo's bale."

Keats.

The poets have celebrated this flower also by its humbler name of Daffodil:

> "Bid Amaranthus all his beauty shed,
> And daffodillies fill their cups with tears,
> To strew the laureat hearse where Lycid lies."
>
> MILTON.

There is a beautiful allusion to the early flowering of the Daffodil in the Winter's Tale:

> ——————————" Daffodils,
> That come before the swallow dares, and take
> The winds of March with beauty."

There is a species of the Daffodil which is very commonly seen by brooks and rivulets with some of the Iris, or Flag-flowers:

> ——————————" there
> Spring the little odorous flowers,
> Violets, and lilies white
> As the slender streams of white
> Gathering about the moon,
> On a lovely eve in June.
> Narcissus hanging down his head,
> And Iris in her watery bed,
> Round about the silver streams,
> Sparkle out like golden beams
> Scattered from Apollo's hair,
> When springing to the morning air
> From the frothy sea, he shook
> Some crystal drops into the brook."

The cup in the centre of the flower is supposed to contain the tears of Narcissus; to which Milton alludes in the passage cited above; and Virgil in the following, where he is speaking of the occupations of the bees:

> ——————" pars intra septa domorum
> Narcissi lacrymam, et lentum de cortice gluten,
> Prima favis ponunt fundamina, deinde tenaces
> Suspendunt ceras."
>
> VIRGIL, GEORGIC 4.

"Some within the house lay tears of daffodils, and tough glue from the barks of trees, for the foundations of the combs, and then suspend the tenacious wax."—MARTYN'S TRANSLATION.

Thomson celebrates the sweetness of the Jonquil, or Sweet Narcissus:

"No gradual bloom is wanting; from the bud,
First-born of Spring, to Summer's musky tribes;
Nor hyacinths, of purest virgin white,
Low bent, and blushing inward; nor jonquils,
Of potent fragrance; nor Narcissus fair,
As o'er the fabled fountain hanging still."

THOMSON'S SPRING.

Virgil, in one passage in the fifth pastoral, speaks of the Narcissus as purple; and Mr. Davidson, in a note on that passage, observes that Dioscorides also speaks of a species of Narcissus which is purple. Several of them have a ring of purple:

"Pro molli violâ, pro purpureo narcisso."

"In lieu of the soft violet, in lieu of the empurpled narcissus."—DAVIDSON'S TRANSLATION.

NASTURTIUM.

TROPÆOLUM.

TROPÆOLEÆ. OCTANDRIA MONOGYNIA.

Called also Indian-cress.—*French,* la capucine.—*Italian,* fior cappucino; caprivola. The botanical name of this plant is the diminutive of *tropæum,* a trophy.

THE Nasturtium is a Peruvian plant; yet, in warm sheltered situations, will grow and flower in the open air, which is extraordinary in a native of so warm a country. They will, however, flower earlier and better when raised in a hot-bed. Where this aid cannot be allowed them, the seed may be sown in autumn, one in a pot; and should be kept in the house till spring. Early in spring they may be gradually inured to the open air. They are esteemed annual plants, but may, with care, be preserved through the winter: they only require protection from frost. There are the Great and the Small Nasturtium, and a double-flowered variety of each. Their colour is a pale yellow, or a deep orange, inclining to red.

The Great Nasturtium, being, from its size, much handsomer than the other, has caused that to be comparatively neglected; and, for a time, it was almost lost to English gardens.

This plant begins to flower in July, and continues till the approach of winter. The blossoms are frequently eaten in salads, and are used for garnishing dishes: the seeds are pickled, under the false name of capers, and, by some persons, are much esteemed. The stalks of the Great Nasturtium will sometimes grow six or eight feet high, and should be trained to some kind of frame for support, if there are several together: where there is only one, a simple stick will suffice.

We are told by Linnæus, that his daughter, Elizabeth Christina, observed the flowers of the Great Nasturtium to emit spontaneously, at certain intervals, sparks, like electric ones, visible only in the evening.

Notwithstanding the glowing and sunny beauty of this well-known flower, it has, I believe, been almost overlooked by those immortal bestowers of immortality, the poets: yet it deserves their attention, no less from the elegance of its foliage, than from the brilliancy of its blossoms, and a certain originality, as it were, in its whole character. Many agreeable things might be said about it, with an allusion, by the way, to the very poetical discovery of Linnæus's daughter. Singular leaves, fire-coloured flowers, a lady, sparks of light, and an evening,—what might not a poet make of all these!

NIGELLA.

RANUNCULACEÆ. POLYANDRIA PENTAGYNIA.

Called also Fennel-flower.—Nigella is a corruption of Nigrella, a name given to this plant from the blackness of its seeds. It is also familiarly called Gith; Bishop's-wort; Devil-in-a-bush; St. Katherine's-flower; Love-in-a-mist.—*French,* la nielle Romaine [Roman nigella]; nielle des jardins [garden nigella]; cumin noir [black cumin]; faux cumin [false cumin]; toute epice [all-spice].—*Italian,* nigella Romana; nigella odorata; melantio; melantro; both from the Greek, and denoting the blackness of its seeds, like the Latin, nigella.

THE kinds of Nigella most esteemed and cultivated in English gardens are the double varieties of the Common, and the Spanish species. They are annual plants: the seeds may be sown in March, three or four in a middle-sized pot. It may stand abroad, and the earth should be kept tolerably moist. It will begin to flower in June or July, and continue till September. The colour varies, but is generally blue or white.

OLEANDER.

NERIUM.

APOCINEÆ. PENTANDRIA MONOGYNIA.

Called also Rose-bay.—*French,* laurier-rose; le laurose commun; le laurose d'Europe.—*Italian,* rosa-lauro; oleandro; nerio; mazza di S. Giuseppe [St. Joseph's staff]; ammazza l'asino [ass bane].

THE Oleanders are nearly allied to the Rhododendrons, but are less hardy: the Common Oleander, indeed, bears the name of Rhododendron also, though not belonging to that genus. It grows by the side of streams, and by the sea-shore, in the Levant, the South of Europe, and in the island of Crete, where it grows very large. In this country its height seldom exceeds eight or ten feet. The colour of the flowers, which, like the Rhododendrons commonly so called, come out in large magnificent bunches at the ends of the branches, varies from purple to a dusky white, a brilliant scarlet, or a deep rich crimson. There are also double-flowered varieties.

This is a beautiful evergreen shrub, requiring the same treatment as the Myrtle; that is, shelter from September to April, a liberal watering every evening in hot weather, and a more sparing draught twice or thrice a week in the winter.

The White-flowered variety is rather more tender, and is usually kept within doors till June.

Most of the Oleanders are East Indians, and require a stove in this country. Some of these are beautiful beyond expression, particularly the Sweet-scented species. The Common Oleander, in addition to the names already mentioned, is also called Rhodo-Daphne. The Hindoos, as we are told by Sir W. Jones, bestow on this handsome shrub a name somewhat less elegant, and most singularly resembling one of its Italian appellations. "They call it," says he, "Horse-killer, from a notion that horses, inadvertently eating of it, are killed by it: most of the species, especially their roots, have strong medicinal, probably narcotic, powers."

OLIVE-TREE.

OLEA.

OLEINEÆ. DIANDRIA MONOGYNIA.

French, l'olivier.—*Italian,* ulivo; olivo; when wild, olivastrello salvatico.

OLIVES are evergreen trees or shrubs, and some of the species are common in drawing-rooms, balconies, &c. In this country they require winter shelter, from September till May, in common seasons. The earth should not be suffered to remain dry, but water should be given in small quantities. The blossoms are white, and very small.

The unripe fruit of the Olive, pickled, of the Provence and Lucca kinds in particular, is to many persons extremely grateful, and is often eaten after dinner with wine: it is supposed to promote digestion, and excite appetite. The oil expressed from the fruit is one of the purest of all the vegetable oils: it is the kind commonly used for culinary purposes; and with the exception of the oil of almonds is the most frequently directed for medicinal preparations.

The Olive is common to all the quarters of the earth; it is celebrated in scriptural history, the dove, which Noah sent out from the ark, returning with an Olive-branch in his bill.

According to poetical history, the Olive was presented to the world by Minerva. We are told that a contest arose between that goddess and Neptune for the right of giving a name to the city of Athens; and that Jupiter decreed that the right should belong to whichever of them should confer the most beneficial gift upon mankind.

"The sea-god stood, and with his trident strake
The cleaving rock, from whence a fountain brake;

Whereon, he grounds his claim. With spear and shield
Herself she arms: her head a murrion steild:
Her breast her Egis guards. Her lance the ground
Appears to strike; and from that pregnant wound
The hoary Olive, charged with fruit, ascends.
The Gods admire: with victory she ends *."

SANDYS'S OVID, Book Sixth.

The more general belief is, that the stroke of Neptune's trident produced a horse. Whichever it may have been, there seems, notwithstanding the great utility of the Olive, to be some ground of suspicion that Minerva owed her victory chiefly to the gallantry of the gods assembled.

The virtues of the Olive, however, are partly emblematical: it is considered as the symbol of peace; and if, in the character of the Goddess of Wisdom, she so far overcame her warlike propensities as to dispose mankind to peace, she cannot be sufficiently honoured for so estimable a benefit.

Spenser tells the story differently, and in a manner more according with the general belief: he describes Minerva as representing the contest in embroidery:

"She made the story of the old debate,
Which she with Neptune did for Athens try;
Twelve gods do sit around in royal state,
And Jove in midst, with awful majesty,
To judge the strife between them stirred late:
Each of the gods by his like visnomy
Eathe to be known, but Jove above them all,
By his great looks and power imperial.

Before them stands the god of seas in place
Claiming that sea-coast city as his right,
And strikes the rocks with his three-forked mace;
Whenceforth issues a warlike steed in sight,
The sign by which he challengeth the place.
That all the gods, which saw his wondrous might,

* "Pliny says the olive-tree, produced on that occasion by Minerva, was to be seen in his time at Athens."

SEE NOTES OF MARTYN'S VIRGIL.

Did surely deem the victory his due:
But seldom seen, forejudgment proveth true.

Then to herself she gives her Ægide shield
And steel-head spear, and morion on her head,
Such as she oft is seen in warlike field:
Then sets she forth how with her weapon dread
She smote the ground, the which straightforth did yield
A fruitful olive-tree, with berries spread,
That all the gods admired; then all the story
She compassed with a wreath of olives hoary."

MUIOPOTMOS.

It was formerly a custom, especially in Athens, for ambassadors to bear an Olive-branch, as an expression of their pacific intentions:

"Yet might they see the Cretans under sail
From high-built walls; when with a leading gale
The Attic ship attained their friendly shore:
Th' Æacides him knew (though many a day
Unseen), embrace, and to the court convey.
The goodly prince, who yet the impression held
Of those perfections which in youth excelled,
Enters the palace, bearing in his hand
A branch of Attic olive."

SANDYS'S OVID, Book Seventh.

Peace is always represented with either a branch or a crown of Olive. Mr. Hunt, in his Mask, expressively twines Myrtle with the Olive of Peace. Milton also puts a Myrtle sprig in her hand:

"But he her fears to cease,
Sent down the meek-eyed Peace;
She, crowned with olive green, came softly sliding
Down through the turning sphere,
His ready harbinger,
With turtle-wing the amorous clouds dividing,
And waving wide her myrtle wand,
She strikes an universal peace through sea and land."

Virgil makes frequent mention of the Olive, and of the situation in which it best thrives:

"Difficiles primum terræ, collesque maligni,
Tenuis ubi argilla! et dumosis calculus arvis,
Palladiâ gaudent silvâ vivacis olivæ.
Indicio est tractu surgens oleaster eodem
Plurimus, et strati baccis silvestribus agri."

VIRGIL, Georgic 2.

"In the first place, stubborn lands, and unfruitful hills, where the bushy fields abound with lean clay and pebbles, rejoice in a wood of long-lived Palladian olives. You may know this soil by wild olives rising thick, and the fields being strewed with wild berries."

MARTYN'S TRANSLATION.

ORANGE-TREE.

CITRUS AURANTIUM.

AURANTIACEÆ. POLYADELPHIA ICOSANDRIA.

The derivation of the word Citrus is unknown: some say it is the name of a place in Asia; others will have it of African origin; some fix it on the Arabian.—*French,* l'oranger.—*Italian,* melarancio; arancio; melangolo.

THE Orange most known in England is the China or Portugal Orange, so called from its having been brought from China by the Portuguese. There are several other varieties in the English gardens; as the Turkey-orange, the Double-flowering, the Dwarf or Nutmeg-orange, the Seville, &c.

The leaves of the Dwarf-orange are very small, and grow in clusters; the flowers grow very close together and appear like a nosegay, the branches being completely covered with them. This species is very ornamental; and, when in blossom, will perfume a room most delightfully. The blossom is white, and begins to appear in June.

Towards the middle of September Orange-trees should be housed; and it would be well to keep them in an inhabited room, but not too near a fire. When it is not frosty, they should be frequently, but sparingly, watered. About April the earth should be removed as deep as can be done without disturbing the roots, and fresh earth supplied. Early in June they may be replaced in the open air; but must be sheltered from keen winds, and from the noon-day sun, which would be hurtful to them. The morning sun will be very beneficial, as also the gentle dews of morning:

> "E quale annunziatrice degli albori
> L 'aura di Maggio muovesi e olezza
> Tutta impregnata dall 'erba e da 'fiori."
>
> DANTE PURGATORIO, 24.

> ——"when to harbinger the dawn, springs up
> On freshen'd wing the air of May, and breathes
> Of fragrance, all impregned with herb and flowers."
>
> CARY'S TRANSLATION.

Every second year the plants should be newly potted at this season; all the roots, on the outside of the ball of earth attached to them, should be cut off; as much of the old earth taken away as can be done without tearing the roots; and the plants set in a tub of water for a quarter of an hour, to soften the lower part of the ball. The stem and leaves should be cleansed with water and a soft woollen cloth. Some stones should be placed at the bottom of the pot, and on these some earth, purposely obtained for Orange plants, to the depth of three or four inches. The plant should then be placed upright in the middle, and the pot filled up, within an inch of the top, with the same earth, being pressed hard down with the hands. The plant should then be watered all over, the watering-pot having the rose on. After this transplanting, the plants should remain in the house a week or two later than on the intervening year, that they may take firm root before they are exposed to the air. In dry summer weather, they should be watered every evening, both roots and leaves; observing to shed water on the leaves, from a rose finely perforated on the spout of the watering-pot. This must not be done until after sunset, or it will cause the leaves to scorch. This caution will apply to plants in general. Water should not be allowed to remain in the saucers: it is injurious to most plants, but to Orange-trees in particular. Another thing to be observed with respect to these plants is, not to put them in pots or boxes too large for them. The largest size used for them should not exceed twenty-four inches in diameter, and much smaller will suffice for the first eight or nine years.

The Seville Orange is the most hardy, and has the largest and most beautiful leaves. The China Orange rarely produces good fruit in England: the varieties with striped leaves never produce it good, nor do they bear so many blossoms as the plain ones.

To have Oranges in perfection, it is considered necessary to graft the trees, even in the warm countries of which they are natives:—"We rode deeper into the wood, and refreshed ourselves with wild Oranges *(laranja da terra)*, which have a mawkish, sweet taste. Oranges, to be good, must be grafted; even in Brazil, if suffered to grow wild, the fruit is flat and rather bitter. Their flowers emitted a delicious smell, and attracted a great number of humming-birds."—PRINCE MAXIMILIAN'S TRAVELS IN BRAZIL, page 76.

In another part of his work, the same author says:—"The heat was intense; we therefore refreshed ourselves with cold punch and excellent Oranges, which in many parts may be had gratis. This excellent fruit can be eaten without injury to the health, even when a person is overheated; but in the evening it is said not to be wholesome. Much more caution is necessary in eating cocoa-nuts and other cooling fruits."—Page 61.

The Brazilians are probably the only people who think

so much caution necessary in eating oranges, as to refrain from their use in the evening.

The following passage may be found in a note in Koster's Brazil:—" Labat says, 'On employe le suc des oranges aigres avec un succès merveilleux et infaillible à guérir les ulcères quelque vieux et opiniâtres qu'ils puissent etre*.' The orange is cut into two pieces, and is rubbed violently upon the wound."—Vol. ii. page 196.

" The first China Orange," says Evelyn, " which appeared in Europe, was sent a present to the old Conde Mellor, then prime minister to the king of Portugal; but of the whole case sent to Lisbon, there was but one plant which escaped the being so spoiled and tainted, that, with great care, it hardly recovered to be since become the parent of all those flourishing trees of that name, cultivated by our gardeners, though not without sensibly degenerating. Receiving this account from the illustrious son of the Conde, I thought fit to mention it for an instance of what industry may produce in less than half an age."

Mickle, in the History of the Portuguese Empire in Asia, prefixed to his translation of the Lusiad, informs us " that the famous John de Castro, the Portuguese conqueror in Asia, was said to have been the first who brought the Orange-tree to Europe, and to have esteemed this gift to his country as the greatest of his actions." He adds, " that Orange-trees are still preserved at Cintra, in memorial of the place where he first planted that valuable fruitage."

The Orange-tree is thought to produce more fruit, if deprived of some of its blossoms. Rapin, in his Poem on Plants, recommends that the nymphs should be allowed, unchecked, to pluck the silvery blossoms, to adorn their bosoms and their vases. " Let your wife, your children,

* They employ the juice of sour oranges with wonderful and infallible success in the cure of ulcers, however old and obstinate.

your whole family be there," says he, "and let them bear away a portion of the fragrant spoils."

The Orange is supposed to be the golden apple presented to Jupiter by Juno on the day of their nuptials. These apples could be preserved nowhere but in the gardens of the Hesperides, where they were protected by three nymphs bearing that name, the daughters of Hesperus; and by a more effectual and appalling guard, a never-sleeping dragon. It was one of the labours of Hercules to obtain some of these golden apples: he succeeded, but, as they could not be preserved elsewhere, it is said they were carried back again by Minerva.

These, too, were the golden apples by means of which Hippomenes won the Arcadian Atalanta; who halted in the race to pick them up, when he artfully dropped them at three several times, in the hope of her so doing: he having received them for that purpose from the goddess Venus.

And probably this may be the golden apple, the bestowal of which first gave origin to the Grecian war.

The Orange-tree is mentioned both by Cowley and Rapin; but the poems being originally written in Latin, and the translations very poor, they will not admit of quotation. It has been celebrated by poets ancient and modern; and well has it deserved its fame, not only for its fine fruit, but also for its handsome leaves, exquisite blossoms, and delicious perfume.

Mr. Moore gives a pleasant picture of the Orange-tree, in his Paradise and the Peri.

> "Just then beneath some orange-trees,
> Whose fruit and blossoms in the breeze
> Were wantoning together, free,
> Like age at play with infancy ——."

The Orange-tree is one of the very few which at once delight us with the promise of spring, and the ripe luxuriance of summer. The poet tells us in his notes, that from the Orange-trees of Kauzeroon the bees cull a celebrated honey.

——————————" in short
All the sweet cups to which the bees resort,
With plots of grass, and perfumed walks between,
Of citron, honeysuckle, and jessamine,
With orange whose warm leaves so finely suit,
And look as if they'd shade a golden fruit."

STORY OF RIMINI.

——————————" thus was this place
A happy rural seat of various view:
Groves whose rich trees wept odorous gums and balm;
Others whose fruit, burnished with golden rind,
Hung amiable, Hesperian fables true,
If true, here only, and of delicious taste.

PARADISE LOST, Book Fourth.

Cowper places the Orange in his green-house:

——————————" The golden boast
Of Portugal and western India there,
The ruddier orange, and the paler lime,
Peep through their polished foliage at the storm,
And seem to smile at what they need not fear."

COWPER'S TASK.

" The garden of Proserpina this hight,
And in the midst thereof a silver seat
With a thick arbour goodly overdight,
In which she often used from open heat
Herself to shroud, and pleasures to entreat.
Next thereunto did grow a goodly tree,
With branches broad dispread, and body great,
Clothed with leaves that none the wood mote see,
And loden all with fruit, as thick as thick might be.

" The fruit were golden apples glistering bright,
That goodly was their glory to behold,
On earth like never grew, ne living wight
Like ever saw, but they from hence were sold,
For those which Hercules with conquest bold
Got from great Atlas' daughters, hence began,
And planted there, did bring forth fruit of gold,
And those with which th' Eubœan young man wan,
Swift Atalanta, when through craft he her outran.

" Here also sprang that goodly golden fruit,
With which Acontius got his lover true,
Whom he had long time sought with fruitless suit,
Here eke that famous golden apple grew,
The which emong the gods false Ate threw
For which th' Idæan ladies disagreed,
Till partial Paris dempt it Venus' due,
And had (of her) fair Helen for his meed,
That many noble Greeks and Trojans made to bleed."

SPENSER'S FAIRY QUEEN.

——————" her lover's genius formed
A glittering fane, where rare and alien plants
Might safely flourish: where the citron sweet
And fragrant orange, rich in fruit and flowers,
Might hang their silver stars, their golden globes,
On the same odorous stem ———."

MASON'S ENGLISH GARDEN.

Mrs. C. Smith speaks of the Orange-tree in her lines addressed to the humming-bird; a beautiful little creature, which, when stript of its plumage, is not bigger than a bee; and, like the bee, it delights in hovering over the sweetest flowers, and sipping their juice, without doing them the least injury by its visit. Mr. Lambert, in his Travels in Canada, says, " that they may be seen there in great numbers, and that their plumage is as beautiful as that of the peacock." It is frequently called the bee bird:

" There, lovely bee-bird ! may 'st thou rove
Through spicy vale, and citron grove,
And woo and win thy fluttering love
With plume so bright;
There rapid fly, more heard than seen,
Mid orange-boughs of polished green,
With glowing fruit, and flowers between
Of purest white."

PEONY.

PÆONIA.

RANUNCULACEÆ. POLYANDRIA DIGYNIA.

From Pæon, an eminent physician of antiquity. It is also a name given both to Apollo and to Esculapius.—*French,* la pivoine; pione: in the village dialect, herbe de mallet; flor de mallet.—*Italian,* rosa de' monti [mountain rose].

THE Peony, from the nature of its roots, requires very deep pots. There are many and beautiful varieties. The White-flowered Peony is a native of Siberia: it is a handsome flower, with the scent of the Narcissus.

The Daurians boil the roots in their broth, and grind the seeds to put into their tea: they call it Dschina.

The Common Peony is purple or red: there are single and double flowers. It is a native of many parts of Europe, of Mount Ida, China, and Japan. A variety which Miller calls the Foreign, Gerarde calls Turkish, and says it originally came from Constantinople. The Portugal variety is a single flower, but very sweet: this requires a lighter soil and a warmer situation than the other kinds. Although the Peony is better adapted for the open ground, it is too beautiful to be dispensed with, where room can be allowed: the Jagged kind is the least fit for pots, and by far the least desirable.

The immense crimson flower of the Double-red Peony is scarcely more magnificent than its luxuriant foliage.

They may be increased by parting the roots, observing to preserve a bud on the crown of each offset, and not to divide them very small: they should be planted three inches deep. It is a hardy plant, and will grow in any soil or situation. They should be kept moderately moist. The Common Peony flowers in May; the White Peony a month later.

PASSION-FLOWER.

PASSIFLORA.

PASSIFLOREÆ. GYNANDRIA PENTANDRIA.

The Passion-flower derives its name from an idea, that all the instruments of Christ's passion are represented in it.—*French,* le grenadille; fleur de la passion.—*Italian,* granadiglia; fiore della passione.

Most of the Passion-flowers are natives of the hottest parts of America, and require a stove in this country. It is a beautiful genus. The rose-coloured Passion-flower is a native of Virginia, and is the species which was first known in Europe. It has since been in great measure superseded by the blue Passion-flower, which is hardy enough to flower in the open air, and makes an elegant tapestry for an unsightly wall. The leaves of this, in the autumn, are of the most brilliant crimson; and, when the sun is shining upon them, seem to transport one to the gardens of Pluto.

The Rose-coloured, however, is better adapted for pots; and, if sheltered from frost, will thrive without artificial heat. In mild weather it may be allowed fresh air, and in the summer will enjoy a full exposure to it. The flowers are purple and white; very handsome, and sweet, but very short-lived; opening in the morning, and fading in the evening.

The fruit is about the size of an Orlean-plumb: when ripe, it is of a pale orange-colour, and encloses many rough seeds, lying in a sweet pulp. The fruit of some kinds is eatable, and in the West Indies much esteemed. It varies in size from that of an olive to that of a large melon.

The fruit of the Laurel-leaved Passion-flower, or Water-lemon, contains a sweet and tasteful juice, which is extremely fragrant. The West Indians suck this juice through a hole in the rind. The French call this species *pomme de liane* [bindweed apple], and English Honeysuckle.

The Passion-flower should be raised in a hot-bed; and should be housed in October, carefully screening it from frost, but admitting air in mild weather. In summer, the earth must be kept tolerably moist, but water must be given very sparingly in winter.

PERWINKLE.

VINCA.

VINCEÆ. PENTANDRIA MONOGYNIA.

French, pervenche; pucellage; violette des sorciers [magician's violet]; vence.—*Italian,* pervinca; centocchio [hundred eyes].

The Perwinkle is a lovely plant: its blue flowers are in bloom all the summer, and its fine glossy green leaves, like large Myrtle-leaves, flourish through the winter. It spreads so fast, and in consequence requires so much room, that it is seldom grown in pots; but it may be preserved very well in that manner, if room can be allowed for it. In a moist soil, and enjoying the morning sun, it thrives and flowers best.

The Madagascar Perwinkle is a beautiful plant, with an upright stem, three or four feet high: the flowers are crimson or peach-coloured on the upper surface, and a pale flesh-colour on the under: it varies with a white flower, having a purple eye. This plant is usually kept in the stove, but the temperature of a warm inhabited room will preserve it very well. Unless the summer prove warm and fine, it must not be set abroad even then; for, if exposed to much wet or cold, it will soon perish. Very little water will suffice this plant.

Chaucer repeatedly mentions the Perwinkle: it makes one of the ornaments of the God of Love:

> "His garment was evèry dele
> Ipurt-raied, and wrought with floures,
> By divers medeling of coloures;
> Floures there were of many gise
> Iset by compace in a sise;
> There lacked no floure to my dome,
> Ne not so moche as floure of brome,
> Ne violet, ne eke pervinke,
> Ne floure none that men can on thinke;
> And many a rose lefe full long
> Was intermedlid there emong;
> And also on his hedde was set
> Of roses redde a chapilet."
>
> THE ROMAUNT OF THE ROSE.

Again in the same poem, the poet, in describing a garden where flowers of all seasons are met together, gives a place to the Perwinkle:

> "There sprange the violet al newe,
> And fresh pervinke, rich of hewe,
> And flouris yelowe, white, and rede;
> Suche plente grewe there ner in mede:
> Ful gaie was all the grounde and queint,
> And poudrid as men had it peint,
> With many a freshe and sondry floure,
> That castin up ful gode savour."

Rousseau has, to his admirers, given the Perwinkle a double interest. He tells us, that walking with Madame Waren, she suddenly exclaimed, "There is the Perwinkle yet in flower." Being too short-sighted to see the plant on the earth without stooping, he had never observed the Perwinkle: he gave it a passing glance, and saw it no more for thirty years. At the end of that period, as he was walking with a friend, "having then begun," he says, "to herborise a little, in looking among the bushes by the way, I uttered a cry of joy: 'Ah, there is the Perwinkle!' and it was so." He gives this as an instance of the vivid recollection he had of every incident occurring at a particular period of his life. The incident is so natural, and told with so much simplicity, that, trifling as it is, it cannot fail to interest; especially as the Perwinkle is in France esteemed as the emblem of sincere friendship, in their mystic language of nosegays, when sent as presents between lovers and friends. The country people in Italy make garlands of it for their dead infants, for which reason they call it *fior di morto* [death's flower].

PHILLYREA.

CASSINE CAPENSIS.

RHAMNEÆ. PENTANDRIA TRYGYNIA.

THIS is an evergreen shrub, bearing white blossoms, which blow in July or August. It should be housed in September, and placed abroad again in May. In the open ground it will thrive well without shelter; and Evelyn says, "is as hardy as the Holly itself." It must be sparingly watered.

This shrub is very similar to the Alaternus, from which it may be distinguished by the position of the leaves; which are opposite on the Phillyrea, alternate in the Alaternus.

The Alaternus is not well adapted for pots, on account of its far-spreading roots; or it would be particularly desirable to a lover of plants, as being one which the tasteful Evelyn prided himself upon bringing into proper notice.

PHLOX.

POLEMONIACEÆ. PENTANDRIA MONOGYNIA.

Called also Lychnidea.

THESE plants are chiefly North American, and most of them tolerably hardy. They are small, the blossoms purple; blowing at different seasons, from April to September, according to their species. They like a moist soil, and must be liberally watered: should be housed in October, and placed abroad again early in May. The usual colour of the blossoms, called by the gardeners purple, is in truth rather a purple-tinged flesh-colour, like that of Venus's Looking-glass. They are numerous and handsome.

In Captain Franklin's Narrative of a Journey to the shores of the Polar Sea, he speaks of a species of Phlox, which, from his unfortunate friend Lieutenant Hood, he names Phlox Hoodii. " This beautiful species," says he, " is a striking ornament to the plains in the neighbourhood of Carlton House, forming large patches, which are conspicuous at a distance."

PINK.

DIANTHUS.

CARYOPHYLLÆ. DECANDRIA DIGYNIA.

The name Dianthus is of Greek origin, and signifies the flower of Jove; which noble name is, according to some, bestowed upon the flower for its beauty; others say from its fragrance. That distinction is surely just, which excites a doubt only for *which* of its good qualities it is conferred.—*French,* oeillet.—*Italian,* garafano; gherosano; garofolo.

The Bearded Pink, or Sweet-William—French, *oeillet de poete*—is a native of Germany. Gerarde mentions it as being, in his time, highly esteemed " to deck up gardens, the bosoms of the beautiful, garlands, and crowns for pleasure." The narrow-leaved kinds are called Sweet-Johns: the broad-leaved, unspotted kinds are by some named Tolmeiners and London-tufts; and the small speckled kind, London-pride.

There are many varieties of the Sweet-William, single

and double-flowered, varying in colour from a pale blush-colour to a deep crimson: some are entirely white. They blow in June: they must be sparingly watered, but never left dry.

Of the Clove-pink—in Italian, *garofano ortense*—the varieties are endless: the larger kinds are called Carnations—in French, *oeillet des fleuristes*:—the smaller, Clove-gilliflowers. Some suppose this latter name to have been corrupted from July-flower, July being its flowering time. Drayton so names it:

> "The curious choice clove July-flower,
> Whose kinds hight the carnation,
> For sweetness of most sovereign power
> Shall help my wreath to fashion;
> Whose sundry colours, of one kind,
> First from one root derived,
> Them in their several suits I'll bind,
> My garland so contrived."
>
> FIFTH NYMPHAL.

It is more generally believed to be from the French name, *giroflier*, which is also the name of the Clove-tree, from the similarity of the perfume: Besides the names already mentioned, Gerarde gives several others, Horse-flesh, Blunket, and Sops-in-wine.

Modern florists have, by their careful culture of these flowers, increased the varieties beyond enumeration. Pinks also, commonly so called, are infinitely varied. Pinks, Carnations, and Sweet-Williams are increased in various ways; but, as their culture demands much attention and experience, the better way of securing handsome varieties will be to purchase them in the pot, which may be done at a small expense; and, as they are chiefly perennial, even that need not often be incurred.

These plants, when in blossom, should be sheltered from the noon-day sun; but suffered to enjoy it in the early part of the day. In dry weather, they should be watered every evening. If the buds on the sides of the stalks are removed, and the top ones only left to blow, they will be much handsomer: should they incline to break through the pod on one side in an unsightly manner, it should be notched in two other places, at equal distances, with a pair of fine scissars, to give them freedom.

When the bloom is past and the leaves decay, the stalks should be cut down; water should then be given but twice a week, observing by no means to use raw spring water. They should be sheltered from frost, and in the spring will again shoot forth in full beauty: they must not be placed very near to a wall, which would tend to draw them up weakly.

The Carnations, Maiden-pinks, &c. have been celebrated both for their beauty and fragrance; in the latter they are equalled by few plants, exceeded perhaps by none. As the rose for her beauty, the nightingale for his song, so is the pink noted for its sweetness.

"And the pink of smell divinest"

is seldom or ever forgotten, when the poets would celebrate the charms of Flora. Spenser's works are continually sprinkled with them: both Milton and Shakspeare have done them honour:

Per. Sir, the year growing ancient,—
Not yet on summer's death, nor on the birth
Of trembling winter,—the fairest flowers o' the season
Are our carnations, and streaked gillyflowers,
Which some call nature's bastards: of that kind
Our rustic garden's barren; and I care not
To get slips of them.
Pol. Wherefore, gentle maiden,
Do you neglect them?
Per. For I have heard it said,
There is an art, which, in their piedness, shares
With great creating nature.

Pol. Say there be,
Yet nature is made better by no mean,
But nature makes that mean.

Winter's Tale, Act 4, Scene 3.

Spenser continually speaks of this flower by the name of Sops-in-wine. Drayton also uses this name for them:

"Sweet-Williams, campions, sops-in-wine,
One by another neatly."

It has been observed that the word Dianthus signifies Jove's flower; but in English the name is generally confined to the Pink, commonly so called; which gives occasion to Cowley to make a facetious remark upon the distinction:

"Sweet-William small has form and aspect bright,
Like that sweet flower that yields great Jove delight;
Had he majestic bulk, he'd now be styled
Jove's flower; and, if my skill is not beguiled,
He was Jove's flower when Jove was but a child.
Take him with many flowers in one conferr'd,
He's worthy Jove, e'en now he has a beard."

Cowley on Plants, Book IV.

POLYANTHUS.

PRIMULA VULGARIS.

PRIMULACEÆ. PENTANDRIA MONOGYNIA.

The Polyanthus bears a great resemblance to the Auricula, and is a variety of Primrose. The roots may be purchased for a trifle, and will live several years: they should be removed and parted every year, and the earth renewed: this may be done in August. The Polyanthus delights in the same rich soil as is recommended for the Auricula, but is a much hardier plant; and needs protection rather from drought and heat, than cold and moisture. It will survive the coldest and the wettest seasons. The Polyanthus, like all of the genus, is an early blower: one of the first flowers which announce spring.

> "Fair-handed Spring unbosoms every grace;
> Throws out the snow-drop and the crocus first;
> The daisy, primrose, violet darkly blue,
> And polyanthus of unnumbered dyes;
> The yellow wallflower stained with iron brown,
> And lavish stock that scents the garden round."
>
> FLOOD OF THESSALY, page 3.

POPPY.

PAPAVER.

PAPAVERACEÆ. POLYANDRIA MONOGYNIA.

Papaver is said to be derived from papa, or pap, because the juice of the flowers was used in pap to produce sleep.—*French*, pavot.—*Italian*, papevero, or papavero, rosoni.

THE red-flowered species, confounded under the name of Corn-poppy, are natives of every part of Europe, the Levant, Japan, &c.; these are with us the most common of all the species, growing in corn-fields, on walls, and on dry banks. They blow in June and July. They are likewise called Red-poppy, Corn-rose, Wind-rose; in Yorkshire, Cup-rose; and in some of the eastern counties, Canker-rose; Red-weed; Head-wark. Gerarde says the country people call them Cheese-bowls. In France, *la pavot rouge des champs; le pavot sauvage; coquelicot; coquelicoq; coque; ponceau; confanon; maudui; graouselle; rouzele;* and in the village dialect, *cabosseta.* In Italy, *papevero erratico; papevero salvatico; rosolaccio.*

The petals of this Poppy give out a fine colour when infused: and a syrup prepared from this infusion is kept in the chemists' shops, but it is not supposed to possess any great medical properties. There is a variety with an oval, black shining spot in the centre: there are likewise some with double flowers, white, red, variegated, &c.; but none are handsomer than the common kind, of a bright scarlet, with a deep purple eye in the centre.

The double flowers, however, are more fit for the present purpose; limited room making it desirable to rear such plants as are of longer duration than the Single-poppy; and these should be sheltered by some tall shrub from the sharp winds, which will otherwise carry them off without mercy.

The Common Black Poppy grows three feet high: it is named from the blackness of its seeds; the flowers are purple. Of this species there are many varieties; some with large double flowers; others variegated with several colours, red and white, purple and white, or finely spotted like some of the Carnations. They are very handsome, but their scent is offensive. They require no shelter, and should be sparingly watered.

The Common White Poppy—called by the French, *le pavot des jardins*; and in Italian, *papavero domestico*—from which chiefly the opium is extracted, is a native of Asia: it grows five or six feet high.

It is chiefly from the seed-vessels of the White Poppy that opium is obtained; but some persons have proposed to substitute the double Red Garden-poppy for the production of this juice; which is come into such frequent use, that the average quantity consumed in Great Britain is no less than 14,400 lb. yearly of Turkey opium. This juice is collected from the White Poppy, grown at about six or eight inches distance from each other, and well watered, until the capsules are half-grown, when the watering is stopped, and the opium is begun to be collected by making at sunset two cuts on the surface of the capsules from below upwards, without penetrating into the cavity, with an instrument that has two points as fine as those of lancets; this is repeated for three or four evenings, when the capsules are then allowed to ripen their seeds. The juice that exudes is collected in the morning, and dried in the sun. An inferior kind of opium is made from the Poppy in the East Indies, and the monopoly of buying it up from the cultivators constitutes the third source of the territorial revenue of the English East India Company, to whom this monopoly produces a million sterling.

Several attempts have been made to collect opium from Poppies grown in England or Scotland. Mr. Young sowed his Poppies for this purpose in April, and found them ready for bleeding in July. The cuts are made by two knives tied together, with guards on their blades that they may not cut deeper than about the sixteenth part of an inch. The juice that exudes is immediately wiped off with a small painter's brush, called by them a sash-tool, rounded a little at the point. When this brush is sufficiently charged, the juice is scraped off by rubbing it on a slip of tin fixed in the mouth of a tin flask. The opium thus collected is then slowly dried without heat, and formed into balls. Mr. Young found that an acre of Poppies thus treated, at five successive bleedings to each head, would yield 56 lb. of opium; and that the Poppy-seeds, on being pressed, yielded 375 pints of salad oil*.

The solution of opium in spirit of wine is now called laudanum, or loddy, so much used instead of tea by the poorer class of females in Manchester and other manufac-

* Transactions of the Society of Arts, vol. xxxvii.

turing towns, and not unknown to the same class in London as a gentle sedative, and the inducer of oblivious delirium from the cares of life. Another preparation of opium, employed, not only for this purpose, but also for quieting the cries of starving children by throwing them into a forced sleep, is that called Godfrey's Cordial, being a coarse syrup made of treacle, flavoured with anise or some similar seed, in which opium is dissolved. Another favourite preparation of this juice is the syrup of poppies, which should be made by boiling the dried capsules (without the seeds) in water, and adding sugar; but as this is a very troublesome process, the syrup is more usually made by dissolving a little opium in a syrup of treacle.

The use of these as stimulants and narcotics, especially in children, without proper care, is highly to be deprecated, and lays in their little frames the foundation of many disorders, besides putting numbers to their last sleep. Opium may be regarded as a gift of heaven itself in some extreme cases, and regulated by the physician; but the danger of its abuse is in proportion.

The author of the Confessions of an English Opium-Eater has so impressively portrayed the fascinations and the terrors of this treacherous drug, and his work has been so popular, that it is unnecessary to enlarge upon the subject here. The reader who takes an interest in it, either will have read, or will choose to read, the book itself.

In Batavia opium is added to tobacco in smoking; a true Dutch improvement.

The Poppy is noted chiefly for its power of inducing sleep, which all the kinds are supposed to possess in some degree.

Thus Virgil, in his Georgics, calls it the Lethæan Poppy, directing it to be offered by way of funeral rite to Orpheus.

Mr. Davidson tells us, in a note to his translation, that it was the custom to offer Poppies to the dead, especially to those whose manes they designed to appease.

Spenser gives it the epithets "dull" and "dead-sleeping:"

"Dull poppy, and drink-quickening setuale."

Speaking of the plants in the Garden of Mammon, he says:

"There mournful cypress grew in greatest store,
And trees of bitter gall, and heben sad,
Dead-sleeping poppy, and black hellebore,
Cold coloquintida,"———

———"not poppy, nor mandragora,
Nor all the drowsy syrups of the world,
Shall ever medicine thee to that sweet sleep
Which thou ow'dst yesterday."

SHAKSPEARE.

"Here henbane, poppy, hemlock here,
Procuring deadly-sleeping;
Which I do minister with fear;
Not fit for each man's keeping."

DRAYTON.

"And thou, by pain and sorrow blest,
Papaver, that an opiate dew
Conceal'st beneath thy scarlet vest,
Contrasting with the corn-flower blue;
Autumnal months behold thy gauzy leaves
Bend in the rustling gale amid the tawny sheaves."

MRS. C. SMITH.

Mr. Hunt, in his Mask, calls it the "Blissful Poppy," from its soothing and sleep-inducing qualities.

————————"O gentle sleep!
Scatter thy drowsiest poppies from above;
And in new dreams, not soon to vanish, bless
My senses with the sight of her I love."

H. SMITH.

"Near the Cimmerians lurks a cave, in steep
And hollow hills, the mansion of dull Sleep:
Not seen by Phœbus when he mounts the skies,
At height, nor stooping; gloomy mists arise
From humid earth, which still a twilight make:
No crested fowle's shrill crowings here awake
The cheerful morn: no barking sentinell
Here guards, nor geese, who wakeful dogs excell:
Beasts tame, nor salvage, no wind-shaken boughs,
Nor strife of jarring tongues, with noises rouse
Secured ease. Yet from the rock a spring,
With streams of Lethe softly murmuring,
Purles on the pebbles, and invites repose:
Before the entry pregnant poppy grows;
With numerous simples, from whose juycie birth
Night gathers sleep, and sheds it on the earth."

SANDYS' OVID.

"Sleep-bringing poppy, by the plowmen late,
Not without cause, to Ceres consecrate:
For being round and full at his half-birth,
It signified the perfect orb of earth;
And by his inequalities when blowne
The earth's low vales and higher hills were showne;
By multitude of grains it held within,
Of men and beasts the number noted bin;
Or cause that seede our elders used to eate,
With honey mixt (and was their after meate);
Or since her daughter that she loved so well,
By him that in th' infernal shades does dwell,
And on the Stygian banks for ever raignes,
(Troubled with horrid cries and noise of chaines)
Fairest Proserpina, was rapt away;
And she in plaints the night, in tears the day,
Had long time spent: when no high power could give her
Any redresse, the poppy did relieve her:
For eating of the seeds, they sleep procured,
And so beguiled those griefs she long endured."

W. BROWNE, vol. ii. 97.

By the ancients the seeds of the White Poppy were served up in their desserts; and they are now used by the Germans to sprinkle over cakes: we use them by their

German name of maw-seed, as a cooling food for singing birds.

The statues of Ceres are commonly adorned with Poppies, they being ever the faithful companions of corn.

Virgil has a fine comparison, which was copied by Ariosto, of a beautiful youth dying, to a Poppy surcharged with rain:

> ——————" Sed viribus ensis adactus
> Transadigit costas, et candida pectora rumpit.
> Volvitur Euryalus leto, pulchrosque per artus
> It cruor, inque humeros cervix collapsa recumbit.
> Purpureus veluti cùm flos succisus aratro
> Languescit moriens; lassove papavera collo
> Demisêre caput, pluviâ cùm fortè gravantur."
>
> VIRGIL, Book IX.

" But the sword, strongly driven, pierces through his side, and rends his white bosom. Euryalus falls to the earth. The blood streams over his beauteous limbs, and his head droops upon his shoulder. Like a purple flower cut down by the plough, he languishes in death; or as a poppy on its weary neck bows down its head when overcharged with rain."

> " Come purpurea fior languendo more,
> Che l' vomere al passar tagliato lassa,
> O come carco di superchio umore
> Il papaver nell' orto il capo abbassa;
> Così, quì della faccia ogni colore
> Cadendo, Dardinel di vita passa;
> Passa di vita, e fa passar con lui
> L' ardire, e la virtù di tutti i sui."
>
> ARIOSTO, Canto 18, Stanza 153.

> " Like the red flower which in its languor lies,
> Left by the plough-share not to rise again;
> Or as the poppy bows its head, and dies
> Beneath the silver burthen of the rain;
> So with his colour fled, and closing eyes,
> Dardinel's soul is gone; he clasps the plain;
> His soul is gone; and with it, gone and fled
> The life and soul of all the men he led."

But Ariosto was not, it seems, the first copyist of this simile; Virgil himself copied it from Homer:

> " As full-blown poppies, overcharged with rain,
> Decline the head, and drooping kiss the plain;
> So sinks the youth: his beauteous head, depress'd
> Beneath his helmet, drops upon his breast."
>
> POPE'S HOMER, Book VIII.

PRIMROSE.

PRIMULA GRANDIFLORA.

PRIMULACEÆ. PENTANDRIA MONOGYNIA.

French, primevère; olive.—*Italian,* prima-vera.

THIS little flower, in itself so fair, shows yet fairer from the early season of its appearance; peeping forth even from the retreating snows of winter: it forms a happy shade of union between the delicate Snow-drop and the flaming Crocus, which also venture forth in the very dawn of spring.

There are many varieties of the Primrose, so called (the Polyanthus and Auricula, though bearing other names, are likewise varieties); but the most common are the Sulphur-coloured and the Lilac. The Lilac Primrose does not equal the other in beauty: we do not often find it wild; it is chiefly known to us as a garden-flower. It is indeed the Sulphur-coloured Primrose which we particularly understand by that name: it is *the* Primrose: it is this which we associate with the cowslips and the meadows: it is this which shines like an earth-star from the grass by the brook-side, lighting the hand to pluck it. We do indeed give the name of Primrose to the Lilac flower, but we do this in courtesy: we feel that it is not the Primrose of our youth;

not the Primrose with which we have played at bo-peep in the woods; not the irresistible Primrose which has so often lured our young feet into the wet grass, and procured us coughs and chidings. There is a sentiment in flowers: there are flowers we cannot look upon, or even hear named, without recurring to something that has an interest in our hearts: such are the Primrose, the Cowslip, the May-flower, the Daisy, &c. &c.

A few Primrose-roots may be transplanted from their native woods or banks; or, should not these be within reach, may be purchased for the value of a few pence at Covent-Garden flower-market. They are perennial; but, being so cheap, it is scarcely worth while to be encumbered with the unsightly roots in winter, when they may be so easily replaced; unless, indeed, we have an individual affection for them, as the gift of a friend, &c.; in such cases they may keep their station, observing now and then to give them a little water, when there is no frost. While in a growing state, they must be plentifully supplied with water, and shaded from the mid-day sun. They like a strong soil, but will thrive in almost any.

The poets have not neglected to pay due honours to this sweet spring-flower, which unites in itself such delicacy of form, colour, and fragrance: they give it a forlorn and pensive character:

"Bring the rathe primrose that forsaken dies."
LYCIDAS.

———————"pale primroses
That die unmarried, ere they can behold
Bright Phœbus in his strength."
WINTER'S TALE.

"The yellow cowslip and the pale primrose."
MILTON'S MAY MORNING.

"What next? a tuft of evening primroses,
O'er which the mind may hover till it dozes;
O'er which it well might take a pleasant sleep,
But that 'tis ever startled by the leap
Of buds into ripe flowers."

KEATS.

"The Primrose, when with sixe leaves gotten grace,
Maids as a true-love in their bosoms place."

W. BROWNE.

The following lines give a pleasant picture of a kind of idly-musing tranquillity:

"As some wayfaring man passing a wood
Goes jogging on, and in his minde nought hath,
But how the primrose finely strew the path,
Or sweetest violets lay downe their heads,
At some tree's roote on mossie featherbeds."

W. BROWNE.

The poems of Clare are as thickly strown with Primroses as the woods themselves; the two following passages are from the Village Minstrel:

"O, who can speak his joys when spring's young morn
From wood and pasture opened on his view,
When tender green buds blush upon the thorn,
And the first primrose dips its leaves in dew!"

"And while he plucked the primrose in its pride,
He pondered o'er its bloom 'tween joy and pain;
And a rude sonnet in its praise he tried,
Where nature's simple way the aid of art supplied."

In another poem, after describing the village children rambling over the fields in search of flowers, he continues:

"I did the same in April time,
And spoilt the daisy's earliest prime;
Robbed every primrose-root I met,
And oft-times got the root to set;
And joyful home each nosegay bore;
And felt—as I shall feel no more*."

* Village Minstrel, &c. vol. i. page 76.

PRIVET.

LIGUSTRUM.

OLEINEÆ. DIANDRIA MONOGYNIA.

In England it was formerly called Prim-print; Prime-print; or Prim.—*French,* troene; fresillon; puîne blanc [white young-one].—*Italian,* ligustro; rovistico; ruistico; olivella: in Venice, conestrela: in the Brescian, cambrosen; cambrosel.

The Privet shrub deserves a place among the most elegant: the leaves are handsome, and the old ones remain on till driven off by new: it bears an abundance of white pyramidal blossoms, which blow in July, and are succeeded by bunches of black berries. It is hardy, and will give little trouble. It must be watered occasionally in dry weather, and must be removed into a roomier lodging, when it has, like the giant in the Castle of Otranto, outgrown its old one.

This elegant tree has been rendered classical by the pen of Virgil:

> "Alba ligustra cadunt, vaccinia nigra leguntur."
>
> Virgil, Pastoral 2.

"White privets fall *neglected,* the purple hyacinths are gathered."—Davidson's Translation.

The Privet blossom has been frequently celebrated for its whiteness:

> "Amarilli, del candido ligustro
> Più candida e più bella,
> Ma dell' aspido sordo
> E più sorda, e più fera, e più fugace."
>
> Guarini, Pastor Fido, Act 1, Scene 2.

> "Amaryllis, yet more fair,
> More white than whitest privets are,
> But than the cruel aspic still
> More cruel, wild, and terrible."

The blossom of the Privet, when exposed to the noonday sun, withers almost as soon as it blows: in the shade it not only lasts longer, but is much larger. The leaves, too, like those of the Laburnum, are much larger and finer when so placed.

PROTEA.

PROTEACEÆ. TETRANDRIA MONOGYNIA.

From Proteus; so named from the great variableness in the fructification.

THE Proteas are elegant shrubs, chiefly natives of the Cape, and requiring protection from our winters. The placing them within doors at that season will generally be sufficient; but care must be taken to water them very sparingly at that time, and to preserve them from damps.

Many of the species are in estimation; among the handsomest are the Grandiflora, the Speciosa, the Cynaroides, the Linearis, the Nana, &c.; the flowers of the latter very much resemble a rose.

RANUNCULUS.

RANUNCULACEÆ. POLYANDRIA POLYGYNIA.

Ranunculus is the diminutive of Rana, a Frog, some of the species growing in the water. It is also familiarly called Gold-cup.

SINCE the introduction of the Persian Ranunculus, the other kinds have been generally neglected; and it has been so much improved by culture, as to vie with the Carnation itself in beautiful varieties. These are of every colour, and combination of colour, that Flora paints with.

As this plant strikes very deep roots, it must be allowed room; though not so much as it will take when in the open ground, where it will often run to the depth of three or four feet. On this account it is better to plant several in one vessel: they may be four or five inches apart, and two deep. The best time to plant them is in October; but, for a succession, they may be continued at intervals even till February. Those first planted will flower in May. When planted in pots, they should be housed in the winter; the roots should be removed every year, cleaned, and dried, and put in a dry place till wanted: they should be removed immediately after the leaves have decayed. A proper soil should be obtained for them, which should be renewed every year. They must be often, but sparingly watered in dry weather.

Many species of the Ranunculus are also familiarly called Crowfoot.

The Aconite-leaved is often cultivated in gardens by the name of White Bachelors'-buttons, or Fair Maid of France; and the Upright Meadow-crowfoot, with double flowers, by the name of Yellow Bachelors'-buttons.

The double-flowered variety of the white ones are very delicate and pretty: they blow in May; the yellow in June and July. They may be increased by parting the roots in autumn.

Shakspeare's Cuckoo-buds of yellow hue are supposed to be the Butter-cup, or King-cup—called by the French, *renoncule; grenouillette; bassinet; pied de coq* [cock's foot]; *pied de corbin* [crowfoot]; *bouton d'or* [gold button]: in the village dialect, *piapau; flor de buro* [butter-flower]: by the Italians, *ranuncolo; boton d'oro; pie corvino*—which belongs to this genus: as also does Wordsworth's Celandine, which has been noticed by that name.

The King-cup is frequently introduced in Clare's poems;

he delights in celebrating wild flowers. It is a curious fact, that notwithstanding the polished beauty of garden flowers, poetry generally prefers to celebrate the wild. The following is a pretty rustic picture:

> "Before the door, with paths untraced,
> The green-sward many a beauty graced;
> And daisy there, and cowslip too,
> And butter-cups of golden hue,
> The children meet as soon as sought,
> And gain their wish as soon as thought;
> Who oft I ween, the children's way,
> Will leap the threshold's bounds to play,
> And, spite of parent's chiding calls,
> Will struggle where the water falls,
> And 'neath the hanging bushes creep
> For violet-bud and primrose-peep.
> And sigh with anxious eager dream
> For water-blobs * amid the stream;
> And up the hill-side turn anon,
> To pick the daisies one by one;
> Then, anxious, to their cottage bound,
> To show the prizes they have found,
> Whose medley flowers, red, white, and blue,
> As well can please their parents too;
> And, as their care and skill contrive,
> In flower-pots many a day survive."
>
> VILLAGE MINSTREL, &c. vol. i. page 76.

He has, in the same volume, another pretty description of flower-gathering, which may find a place here:

> "Some went searching by the wood,
> Peeping 'neath the weaving thorn,
> Where the pouch-leaved cuckoo-bud †
> From its snug retreat was torn.

* Marsh Marygolds.

† Clare's cuckoo-buds are neither the lady's-smock nor the king-cup; neither does he mean the ragged-robin, for that is here expressly distinguished from them: probably he means the arum, or lords and ladies.

Where the ragged robin stood
  With its piped stem streaked with jet,
And the crow-flowers, golden-hued,
  Careless plenty easier met."

Page 137.

In his descriptions of Rural Life and Scenery, he gives this flower for a goblet to the fairies:

" And fairies now, no doubt, unseen
  In silent revels, sup,
With dew-drop bumpers toast their queen
  From crow-flower's golden cup."

RHODODENDRON.

RHODODENDRUM.

RHODORACEÆ. DECANDRIA MONOGYNIA.

This name is of Greek origin, and signifies Rose-tree.

THE Rhododendrons are handsome flowering shrubs, hardy enough to bear the open air in this country. When in pots, however, if the winter be severe, it may be well to cover the roots with a little moss; and some keep them always so covered, to shelter them from frost in the winter, and to preserve the earth moist in the summer. The kind most commonly cultivated here is a native of North America. The flowers are generally rose-coloured or purple; there is a variety with yellow, and one with white flowers.

There is a shrub called Rhodora, a native of Newfoundland, very similar to this, which may be treated in the same manner.

The best time for transplanting these shrubs, when they require removal into a larger pot, is in September or April; and when they do not need this removal, it will be well, at that season, to renew the earth as far as can be done without disturbing the roots; or even occasionally to take them out of the pot, and pare away the decayed roots on the outside of the ball of earth adhering to them.

Mr. Moore quotes a passage from Tournefort, in his notes to Lalla Rookh, informing us that about Trebizond there is a kind of Rhododendron, on the flowers of which the bees feed, and that their honey drives people mad:

> " E'en as those bees of Trebizond,—
> Which from the sunniest flowers that glad
> With their pure smile the gardens round,
> Draw venom forth that drives men mad."

ROBINIA.

LEGUMINOSÆ. DIADELPHIA DECANDRIA.

The Rose-acacia, so called from the colour of its blossoms, is a beautiful shrub: it is a native of Carolina, where it will grow twenty feet high: here it does not exceed six or eight, and will produce flowers when not more than a foot high: the flowers are large, showy, and numerous, and the shrub is of ready growth. It is not very nice as to soil and situation, but prefers a light, moist soil, and a situation rather sheltered than exposed. It blossoms in June.

The Siberian species of Robinia have mostly yellow flowers. The Salt-tree Robinia, of which the blossoms are purple, is an exception. This, and the Thorny Robinia are, at their full growth, about six feet high; the Shrubby, ten feet; the Shining, five; the Daurian and the Dwarf, three feet.

The Salt-tree grows naturally in salt fields, and will

not flower but in a saline soil. These shrubs will bear this climate very well, with the exception of the Rose-acacia, which must be sheltered from frost and keen winds. They may be kept moderately moist.

ROCKET.

HESPERIS.

CRUCIFERÆ. TETRADYNAMIA SILIQUOSÆ.

Hesperis is from the Greek, and signifies evening; the flowers are so called because they smell sweetest at that time.—*English,* Rocket; Dame's-violets; Damask-violets; Queen's-gillowflowers; Rogue's-gilliflowers; Winter-gilliflowers; and Close-sciences.—*French,* la Julienne; la Juliana; la cassolette [smelling-bottle]; la giroflée musquée [musk pink ; la giroflée des dames [ladies' pink]; la violette des dames [dame's violet]; la Juliane de nuit [night Juliana].—*Italian,* esperide; Giuliana; viola matronale [housewives' violet]; bella Giulia [pretty Julia].

The species called the Night-smelling Rocket is much cultivated for the evening fragrance of the flowers, which induces the ladies in Germany to keep it in pots in their apartments, whence it obtained the name of Dame's-violets.

These flowers are generally biennial, and flower but once; they must therefore be frequently supplied. A strong root of each kind desired should be set apart, not suffered to flower; but when the flower-stems have shot up six inches high, they should be cut down close to the bottom: these stalks may each be cut into two pieces, and both halves planted in a soft loamy earth and placed where they may enjoy the morning sun. They should then be well-watered and covered with glasses, round the rims of which the earth should be drawn close, to exclude the air. When the sun is hot, these glasses should be shaded. Once a week the cuttings should be watered, and again carefully covered. With this management they will put out roots in five or six weeks, and begin to shoot above: then the glasses should be a little raised on one side, to admit the air, and gradually to harden them. When they have taken good root, replant them in pots about ten inches in diameter, observing to shade them till they have taken good root, and to water them as when first planted.

The roots so cut down will send out more stalks than before, and these may be cut down and treated in the same manner; so that, if the roots are sound, two or three crops of cuttings may be taken from them, and there may always be a good supply of these flowers.

They blow in June; and, after the flowers have decayed, young plants may be raised from the stalks as before directed; but not so strong as from the fresh roots, nor are they always sure to grow.

Their colours are purple or white; single and double of each; they must not be over-liberally watered, nor planted in a very rich soil, or they will be liable to rot. In dry, hot weather, when they are in flower, they may be watered every evening, but it must be very sparingly.

This beautiful plant is rather scarce in this country, as the cuttings treated in the ordinary way do not succeed well; but the following method will be found a never-failing method of propagating it. After the flower has begun to fade, cut down the stalks and divide them into cuttings, strip off the leaves and smooth the ends, then make three slits with a knife in the rind lengthways, so as to raise it for about half an inch in length. By this means, when the cutting is inserted into the ground, the loose rind curls up, and thus a greater tendency to throw out roots is produced, so that not one in twenty will fail. The same method is equally efficacious in cuttings of stock-gilliflowers and double wall-flowers.

ROSE-BUSH.

ROSA.

ROSACEÆ. ICOSANDRIA POLYGYNIA.

French, le rosier; flowers, la rose: in bas Breton, rôs.—*Italian*, rosajo: rosa: in the Brescian, larrosa.

It is not intended to set down here a catalogue of the various kinds of Roses, but to speak of a few of the most eminent, and particularly such as are best adapted for the present purpose. Unfortunately it happens with many of them, as with some other valuable plants, of which the Laurustinus is one, that they will not thrive well in the vicinity of London.

The Single Yellow Rose is a native of Germany, Italy, and the South of France. The Austrian Rose is considered as a variety of this: it is of a sulphur-colour outside, and a bright scarlet within. The Double Yellow Rose is full and large, as the Provins Rose: it is a native of the Levant.

These kinds are principally mentioned for their rarity, being some of those which will not grow near London.

The Cinnamon Rose—French, *rose canelle*—is one of the smallest and earliest of the double garden roses: it is supposed to be named from the scent of the leaves, some say of the flowers. Mr. Martyn says he can discover nothing in the scent of either, at all resembling that spice.

The Dog-Rose is well known as the blossom of the Common Briars, growing wild in almost every part of Europe; here called the hip-tree, hep-tree, and in Devonshire, canker, and canker-rose: the name of dog-rose probably arises from the heps or fruit being eaten by dogs, whence the Tartars call the heps by a name signifying dog-fruit. In French these roses are called *rosier sauvage*, wild rose-bush; *rosier des haies*, hedge rose-bush; *rose de chien*, dog rose; *rose cochonniere*, swine rose; *eglantier; eglantine:* in Italian they are called *rosa salvatica; rosa canina.*

The Scotch Rose is also common to most parts of Europe; the petals are white, or cream-coloured; yellow at the base and sometimes striped with red: the fruit is a dark purple, and the pericarp contains a fine purple juice, which, diluted with water, dyes silk and muslin peach-colour: the addition of alum will make it a deep violet dye. The fruit, when ripe, is eaten by children: the leaves are small and elegant: the whole plant seldom exceeds a foot in height: it likes the shade and a moist soil.

The Common Provins Rose—French, *rose de Provins*—is one of the most beautiful yet known in the English gardens: it is very large and full, folded close in the manner of a cabbage; some call it the Cabbage-Rose on this account. It is the most fragrant as well as the handsomest kind we have: it will grow seven or eight feet high. The petals, which are deep red and of a powerful scent, may be kept for a year or eighteen months by being pressed close. It takes its name not from Provence, as is commonly supposed, but from Provins, a small town about fifty miles from Paris, where it is largely cultivated: and where it was first introduced from the east.

There are two small varieties of the Provins which are much esteemed, the Rose de Meaux and the Pompone Rose: if the old wood of these kinds be cut down every year after they have done blowing, it will cause them to shoot more vigorously, and to flower more freely.

The Moss-rose, or Moss Provins-rose, is well known as an elegant plant; the flowers are deeply coloured, and the rich mossiness which surrounds them gives them a luxuriant appearance not easily described; but it is familiar to every one. It is a fragrant flower: its country is not known to us, and we know it only as a double flower.

The Red Provins-rose is smaller than the Common Provins, and deeper-coloured; there is also a Blush, and a White Provins.

The Damask-rose is a pale red: it is not very double, but is sweet-scented, and extremely handsome. It is a native of the South of France: there are many varieties, the Monthly, the Striped Monthly, which is red and blush-coloured, and the York and Lancaster, so called because it is striped with both red and white. Miller believes this Rose to have been brought originally from Asia; a syrup is prepared from it.

The Frankfort-rose is full and handsome, but scentless. This and the Damask-rose grow about the same height as the Provins.

The Monthly Roses do not thrive well near London, but are not so peremptory in this point as the Yellow Roses, which it is said will not flower within ten miles of it. Of the other kinds which have been mentioned, the dead wood should be cut out every year, and the suckers taken off: this should be done in the autumn.

The Red Rose is large, but not very double; it is of a rich crimson colour, and particularly fragrant. Parkinson calls this the English Rose, because the first known in this country, and more cultivated here than elsewhere; and because it was assumed by some of our kings as a symbol of royalty. There is a variety of this kind, with white and red stripes. This rose is used in medicine for conserves, infusions, honeys, syrups, &c., and was much valued by Arabian physicians.

Gerarde says, "that in Leylande fields, in Lancashire, this Garden-rose doth grow wild in the ploughed fields among the corn in such abundance, that many bushels of them may be gathered there, equal with the best Garden-roses in every respect; but what is yet more surprising," continues he, "is that in one of the fields, called Glover's field, every year that it is ploughed for corn, it will be spread over with roses; but when not ploughed, then there shall be but few roses to be gathered:" and this he has "by the relation of a curious gentleman there dwelling."

"I give this improbable tale," says Mr. Martyn, after quoting this passage, "as an instance of the dependence that is to be placed upon the information of curious gentlemen." Johnson has set it right by informing us, he had heard that the roses which grow in such plenty in Glover's-field are no other than the *Corn-rose*, or Red Poppy.

The Hundred-leaved-rose is a native of China: they are very double, deeply-coloured, with little scent. This is a most beautiful species: the varieties are numerous; it is often confounded with the Damask-rose, from which it is quite distinct. This Rose is used in medicine, and a fine distilled water of an exquisite perfume is prepared from it; but the oil, or rather butter, that swims on the water has no scent. The water which is prepared from the Common Dog-rose is by many considered as more fragrant than when distilled from any of the Garden-roses. The leaves, too, of this wild kind are used as a substitute for tea; and the fruit when ripe, and mellowed by the frost, is often eaten, and thought very agreeable: it is a great delicacy to some kinds of birds, to pheasants in particular: it is also mixed with sugar and sold under the name of conserve of heps, and forms a good vehicle for many nauseous medicines.

The Evergreen-rose is a native of the South of Europe: it is white, small, single, but very sweet: in appearance it much resembles our Eglantine. It is this rose that yields the fine scented oil called attar of roses, which is imported from the Barbary coast, Egypt, and the East Indies; a few drops of this oil, dissolved in spirit of wine, form the *esprit de rose* of the perfumers; and the same, dissolved in fine sweet oil, their *huile antique de rose.*

The Eglantine, or Sweet-briar-rose, is a native of all Europe, in woods, thickets, hedges, &c., chiefly in a gravelly soil. The varieties with double flowers are very elegant shrubs.

The Musk-rose is common in every hedge in Tunis: it is white, smells strongly of musk, and blows in August: there are single and double varieties. This Rose requires plenty of room.

The Red China Rose is semi-double: it is admired for its fine rich crimson colour, and for its fragrance. It blows in succession all the year, but more sparingly in the winter months. There is also a Blush, and a Pale China Rose.

The White Rose is a native of China and most parts of Europe: it grows to a height of nine or ten feet, is very full of blossom, and extremely beautiful, but has little or no smell.

Roses in general delight in an open free air, and will bear the cold well; but, when in pots, it is better to place them in-doors during the winter, particularly such as flower at that season. The earth should be always kept moderately moist.

The Rose is pre-eminently the flower of Love and Poetry, the very perfection of floral realities. Imagination may have flattered herself that her power could form a more perfect beauty; but, it is said, she never yet discovered such to mortal eyes. This, however, she would persuade us to be a mere matter of delicacy, and that she had the authority of Apollo for her secret success:

——————"no mortal eye can reach the flowers,
And 'tis right just, for well Apollo knows,
'Twould make the poet quarrel with the rose."

It is however determined, that until the claim of such veiled beauty, or beauties, shall rest upon better foundation, the Rose shall still be considered as the unrivalled Queen of Flowers.

> "I saw the sweetest flower wild nature yields,
> A fresh-blown musk-rose."

The Rose, as well as the Myrtle, is considered as sacred to the Goddess of Beauty. Berkeley, in his Utopia, describes lovers as declaring their passion by presenting to the fair beloved a rose-bud just beginning to open; if the lady accepted and wore the bud, she was supposed to favour his pretensions. As time increased the lover's affection, he followed up the first present by that of a half-blown rose, which was again succeeded by one full-blown; and if the lady wore this last, she was considered as engaged for life *.

In our country, in some parts of Surrey in particular, it was the custom, in the time of Evelyn, to plant roses round the graves of lovers †. The Greeks and Romans observed this practice so religiously, that it is often found annexed as a codicil to their wills, as appears by an old inscription at Ravenna, and another at Milan, by which roses are ordered to be yearly strewed and planted upon the graves.

It is the universal practice in South Wales to strew roses and other flowers over the graves of departed friends.

Morestellus cites an epitaph, in which Publia Cornelia Anna declares that she had resolved not to survive her husband in desolate widowhood, but had voluntarily shut herself up in his sepulchre, still to remain with him with whom she had lived twenty years in peace and happiness: and then orders her freed-men and freed-women to sacri-

* See Gaudentio di Lucca.

† Evelyn's Sylva.

fice there to Pluto and Proserpine, to adorn the sepulchre with roses, and to feast upon the remainder of the sacrafice.

We have seen, within these few years, the body of a child carried to a country church for burial, by young girls dressed in white, each carrying a rose in her hand.

Poetry is lavish of roses; it heaps them into beds, weaves them into crowns, twines them into arbours, forges them into chains, adorns with them the goblet used in the festivals of Bacchus, plants them in the bosom of beauty. —Nay, not only delights to bring in the rose itself upon every occasion, but seizes each particular beauty it possesses as an object of comparison with the loveliest works of nature;—As soft as a Rose leaf; As sweet as a Rose; Rosy-clouds; Rosy-cheeks; Rosy-lips; Rosy-blushes; Rosy-dawns, &c. &c. It is commonly united with the lily:

"A bed of lilies flower upon her cheek,
And in the midst was set a circling rose."

P. FLETCHER.

"Rosed all in lovely crimson are thy cheeks,
Where beauties indeflourishing abide,
And as to pass his fellow either seeks,
Seem both to blush at one another's pride."

G. FLETCHER.

"Tell me have ye seen her angel-like face,
Like Phœbe fair?
Her heavenly 'haviour, her princely grace
Can you well compare?
The red-rose medled with the white y-fere,
In either cheek depeinten lively chear:
Her modest eye,
Her majesty,
Where have you seen the like but here?"

SPENSER.

"The rois knoppis, te tand furth thare hede,
Gan chyp, and kyth thare vernale lippis rede;

Crysp skarlet levis sum scheddand baith attanis,
Kest fragrant smel amyd fra goldin granis."

GAWIN DOUGLAS.

" Its velvet lips the bashful rose begun
To show, and catch the kisses of the sun;
Some fuller blown, their crimson honors shed;
Sweet smelt the golden chives that graced their head."

MODERNISED BY FAWKES.

" Had I a cheek like Rhodope's
In midst of which doth stand
A grove of roses such as these,
In such a snowy land:
I would make the lily, which we now
So much for whiteness name;
As drooping down the head to bow,
And die for very shame."

DRAYTON.

" A stream of tears upon her faire cheeks flowes,
As morning dewe upon the damask-rose,
Or crystal-glasse vailing vermilion;
Or drops of milk on the carnation."

W. BROWNE'S PASTORALS.

————" cui plurimus ignem
Subjecit rubor, et calefacta per ora cucurrit.
Indum sanguineo veluti violaverit ostro
Si quis ebur; aut mixta rubent ubi lilia multâ
Alba rosâ; tales virgo dabat ore colores."

VIRGIL, ÆNEID, book xii.

"—Lavinia, in whom profound modesty lighted up a burning flush, and diffused it over her inflamed face. As if one had stained the Indian ivory with ruddy purple; or as white lilies mingled with copious roses blush: such colors the virgin in her image showed."— DAVIDSON'S TRANSLATION.

" The lady lily paler than the moon,
And roses, laden with the breath of June."

BARRY. CORNWALL.

" Sometimes upon her forehead they behold
A thousand graces masking with delight,
Sometimes within her eye-lids they unfold
Ten thousand sweet belgards, which to their sight
Do seem like twinkling stars in frosty night:

But on her lips, like rosie buds in May,
So many millions of chaste pleasures play.

SPENSER.

"The rose, the flower of love,
Mingle with our quaffing;
The rose, the lovely leaved,
Round our brows be weaved,
Genially laughing.

"O the rose, the first of flowers,
Darling of the early bowers,
E'en the gods for thee have places;
Thee too Cytherea's boy
Weaves about his locks for joy,
Dancing with the graces.

"Crown me then; I'll play the lyre,
Bacchus, underneath thy shade:
Heap me, heap me, higher and higher,
And I'll lead a dance of fire
With a dark deep bosomed maid."

HUNT, *from Anacreon.*

"Her face so fair, like flesh it seemed not,
But heavenly portrait of bright angel's hue
Clear as the sky withouten blame or blot,
Through goodly mixture of complexions due;
And in her cheeks the vermeil red did shew
Like roses in a bed of lilies shed,
The which ambrosial odours from them threw,
And gazer's sense with double pleasure fed,
Able to heal the sick, and to revive the dead."

SPENSER.

The Red-Rose is said to have been indebted for its color to the blood which flowed from the thorn-wounded feet of Venus, when running through the woods in despair for the loss of Adonis: as the White-Rose is also said to have sprung from the tears which the goddess shed upon that occasion. Ample reasons these for dedicating them to her.

"White as the native rose before the change,
Which Venus' blood did in her leaves impress."

SPENSER.

So universally as the Rose has been celebrated in full blown beauty, few have done justice to its infant loveliness.

> "Of the rose full lipped and warm,
> Round about whose riper form
> Her slender virgin train are seen
> In their close fit caps of green."
>
> DESCENT OF LIBERTY.

> "And they that set at ease
> The sheath-enfolded fans of rosy bushes,
> Ready against their blushes."
>
> NYMPHS.

These two last lines seem to bring the opening leaves of the Rose-bush immediately before our eyes.

> 'Ah! see the virgin rose, how sweetly she
> Doth first peep forth with bashful modesty,
> That fairer seems, the less ye see her may;
> Lo! see soon after, how more bold and free
> Her bared bosom she doth broad display;
> Lo! see soon after, how she fades and falls away."
>
> SPENSER.

Perhaps the most beautiful season of the Rose is when partly blown; then too she still promises us a continuance of delight; but when full-blown, she inspires us with the fear of losing her. The following lines refer to a Rose plucked from its stem.

> "Look as the flower which lingeringly doth fade,
> The morning's darling late, the summer's queen;
> Spoiled of that juice which kept it fresh and green,
> As high as it did raise, bows low the head."
>
> DRUMMOND.

No true poet can describe a garden, or a bouquet, without telling us that—

> "There was the pouting rose, both red and white."

Apollo would no longer acknowledge him if he overlooked this flower.

The reader will remember that the Red-Rose has from its long dwelling with us been named the English-Rose; doubtless it is to this flower Brown alludes in the following lines, where speaking of the rivulets, he says,

"Some running through the meadows, with them bring
Cowslip and mint: and 'tis another's lot
To light upon some gardener's curious knot
Whence she upon her breast (Love's sweet repose)
Doth bring the queen of flowers, the English-rose."

The bed of roses is not altogether a fiction. "The Roses of the Sinan Nile, or garden of the Nile, attached to the Emperor of Morocco's palace, are unequalled; and mattresses are made of their leaves, for men of rank to recline upon*."

The Eastern Poets have united the Rose with the nightingale; the Venus of flowers with the Apollo of birds: the Rose is supposed to burst forth from its bud at the song of the nightingale.

"You may place a handful of fragrant herbs and flowers before the nightingale; yet he wishes not in his constant heart for more than the sweet breath of his beloved Rose."—Jami †.

"Oh! sooner shall the rose of May
Mistake her own sweet nightingale,
And to some meaner minstrel's lay
Open her bosom's glowing veil,
Than Love shall ever doubt a tone,
A breath of the beloved one."

————"though rich the spot
With every flower this earth has got,
What is it to the nightingale
If there his darling rose is not?"

Lalla Rookh.

A festival is held in Persia, called the feast of Roses, which lasts the whole time they are in bloom.

* See notes to Moore's Lalla Rookh. † Ibid.

"And all is ecstasy, for now
The valley holds its feast of roses;
That joyous time, when pleasures pour
Profusely round, and in their shower
Hearts open, like the season's rose,—
The flowret of a hundred leaves,
Expanding while the dew-fall flows,
And every leaf its balm receives!"

LALLA ROOKH.

Persia is the very land of Roses:

"On my first entering this bower of fairy land," say Sir Robert Kerr Porter, speaking of the garden of one of the royal palaces of Persia, "I was struck with the appearance of two rose-trees full fourteen feet high, laden with thousands of flowers, in every degree of expansion, and of a bloom and delicacy of scent that imbued the whole atmosphere with exquisite perfume. Indeed, I believe that in no country in the world does the rose grow in such perfection as in Persia; in no country is it so cultivated and prized by the natives. Their gardens and courts are crowded by its plants, their rooms ornamented with vases filled with its gathered bunches, and every bath strewed with the full-blown flowers plucked with the ever replenished stems. . . . . But in this delicious garden of Negaaristan, the eye and the smell are not the only senses regaled by the presence of the rose. The ear is enchanted by the wild and beautiful notes of multitudes of nightingales, whose warblings seem to increase in melody and softness with the unfolding of their favorite flowers. Here indeed the stranger is more powerfully reminded, that he is in the genuine country of the nightingale and the rose."—PERSIA IN MINIATURE, vol. 3.

Lord Byron has taken advantage of the various fictions and customs connected with the Rose; and has made it spring and flourish over the tomb of Zuleika: while the nightingale soothes his beloved with his sweet and plaintive notes:

"A single rose is shedding there
Its lonely lustre meek and pale:
It looks as planted by despair—
So white, so faint—the slightest gale
Might whirl the leaves on high;
And yet though storms and blight assail,

And hands more rude than wintry sky
  May wring it from the stem—in vain—
  To-morrow sees it bloom again!
The stalk some spirit gently rears,
And waters with celestial tears;
  For well may maids of Helle deem
That this can be no earthly flower,
Which mocks the tempest's withering hour,
And buds unsheltered by a bower;
Nor droops though spring refuse her shower,
  Nor wooes the summer beam:
To it the livelong night there sings
A bird unseen, but not remote:
Invisible his airy wings,
But soft as harp that Houri strings
  His long entrancing note."

Bride of Abydos.

There is in this poem another passage on the same subject; a passage which instantly brings before our eyes that lovely design by Stothard, of the kneeling Zuleika:

"She saw in curious order set
  The fairest flowers of eastern land—
He loved them once; may touch them yet,
  If offered by Zuleika's hand.

"The childish thought was hardly breathed
Before the rose was plucked and wreathed;
The next fond moment saw her seat
Her fairy form at Selim's feet:
  This rose to calm my brother's fears
  A message from the Bulbul * bears;
  It says to-night he will prolong
  For Selim's ear his sweetest song;
  And though his note is somewhat sad,
  He'll try for once a strain more glad,
  With some faint hope his altered lay
  May sing these gloomy thoughts away."

Some suppose that Syria takes its name from Suri, a beautiful species of Rose, for which that country has been always famous †.

* The nightingale.

† See notes to Lalla Rookh.

"And if at times a transient breeze
Break the blue crystal of the seas,
Or sweep one blossom from the trees,
How welcome is each gentle air
That wakes and wafts the odours there!
For there—the Rose o'er crag or vale,
Sultana of the Nightingale,
 The maid for whom his melody,
 His thousand songs are heard on high,
Blooms blushing to her lover's tale:
His queen, the garden queen, his Rose,
Unbent by winds, unchill'd by snows,
Far from the winters of the west,
By every breeze and season blest,
Returns the sweets by nature given
In softest incense back to heaven;
And grateful yields that smiling sky,
Her fairest hue, and fragrant sigh."

LORD BYRON'S GIAOUR.

We must not dismiss the subject of the Rose, without recalling to the minds of our readers those beautiful lines from Milton:

————————" Eve separate he spies,
Veiled in a cloud of fragrance, where she stood,
Half spied, so thick the roses blushing round
About her glowed; oft stooping to support
Each flower of tender stalk, whose head, though gay
Carnation, purple, azure, or speck'd with gold,
Hung drooping unsustained; them she upstays
Gently with myrtle band, mindless the while
Herself, though fairest unsupported flower,
From her best prop so far, and storm so nigh."

Chaucer delights in garlanding his heads with roses: not the daisy itself delights him more than a garland of flowers; Roses in particular. In the Flower and the Leaf he has crowns of Roses, laurel, oak, woodbine, &c. and of all various flowers mingled together, he perfectly revels in them:

"And all they werin, aftir ther degrees
Chappèlets new, or made of laurir green;

Or some of oke, or some of othir trees;
Some in their hondis havin boughis shene,
Some of laurin, and some of okis bene,
Some of hawthorne, and some of the wodebind,
And many mo which I have not in mind.

"And everich had a chapelet on her hed,
(Which did right wele upon the shining here)
Makid of goodly flouris white and red,
The knightis eke that they in hondè led
In sute of them ware chaplets everichone;
And before them went minstrels many one.

"As harpis, pipis, lutis, and sautry,
Allè in grene, and ther hedis bare
Of diverse flouris made full craftily,
All in a sute, godely chaplets they ware,
And so dauncing into the mede they fare,
In mid the which they found a tuft that was
All ovirsprad with flowris in compas."

In two different poems where Venus is represented, she has a crown of white and red flowers:

"I saw anone right her figure
Nakid yfletyng in a se,
And also on her hedde parde
Her rosy garland white and redde."

In the Knight's-Tale he again describes the goddess floating in the sea;

"And on hire hed, ful semely for to see
A rose gerlond fresh and wel-smelling
Above hire hed hire doves fleckering."

"She gathereth floures, partie white and red,
To make a sotel gerlond for hire hed."

St. Cecilia receives a miraculous crown of Roses and lilies; and Cupid is crowned with Roses. In the Romaunt of the Rose, he describes himself as selecting from many Roses the one which shall best please him:

"Of knoppis close some sawe I there,
And some well better woxin were,
And some there ben of othir moison,
That drowe nigh to ther seson,
And spedde 'hem fastè for to spredde:
I lovè well suche rosis redde,
For brode rosis and open also
Ben passid in a day or two,
But knoppis wollin freshè be
Two daies at lest, or ellis thre;
The knoppis gretely likid me,
For fairer maie there no man se;
Who so might hav in one of all,
It ought him ben ful lefe withall:
Might I garlonde of 'hem getten
For no richesse I wolde it letten.

"Amonges the knoppis I chese one
So faire, that of the remenaunt none
Ne preise I half so well as it
Whan I avisin in my wit.
It so well was enluminid
With color red, as well finid
As nature couth it makin faire,
And it hath levis wel foure paire
That kind hath set through his knowing:
About the redde rosis springing
The stalke was as rishè right,
And thereon stood the knop upright,
That it ne bowed upon no side:
The sote smel ysprong se wide
That it died all the place about.

"I sawe the rose when I was nigh,
Was greater woxin, and more high,
Freshe and roddy and faire of hewe,
Of colour ever iliche newe:
And when I had it longè sene,
I sawe that through the levis grene,
The rose spred to spannishing,
To sene it was a godely thing;
But it ne was so sprede on brede
That men within might knowe the sede,
For it coverte ywas and close
Bothe with the leves, and with the rose;

The stalke was even and grene upright,
It was thereon a godely sight."

The short-lived beauty of the Rose has given rise to many reflections and comparisons; as in Crashaw's lines on the death of Mr. Herrys: an instance occurs also in Mr. Bowring's translation from the Russian of Kostrov:

"The rose is my favorite flower:
On its tablets of crimson I swore,
That up to my last living hour
I never would think of thee more.

"I scarcely the record had made,
Ere Zephyr, in frolicsome play,
On his light airy pinions conveyed
Both tablets and promise away."

Bowring's Russian Anthology.

And a beautiful one in Tasso:

"Deh mira, egli cantò spuntar la rosa
Dal verde suo modesta e verginella,
Che mezzo aperta ancora e mezzo ascosa,
Quanto si mostra men, tanto è più bella.
Ecco poi nudo il sen già baldanzosa
Dispiega: ecco poi langue, e non par quella;
Quella non par, che desiata avanti
Fu da mille donzelle e mille amanti.

"Cosi trapassà al trapassar d'un giorno
Della vita mortale il fiore, e 'l verde."

La Gerusalemma Liberata di Tasso: *Canto* 16.

"The gentle budding rose, quoth she, behold,
That first scant peeping forth with virgin beams,
Half ope, half shut, her beauties doth upfold
In its fair leaves, and, less seen fairer seems;
And after spreads them forth more broad and bold,
Then languisheth, and dies in last extremes:
Nor seems the same, that decked bed and bower
Of many a lady late, and paramour:

"So, in the passing of a day, doth pass
The bud and blossom of the life of man."

Fairfax's Translation.

"Tum pater Anchises magnum cratera coronâ
Induit, implevitque mero."

Virgil, Eneid, book 3.

"Then father Anchises decked a capacious bowl with a garland, and filled it up with wine."—DAVIDSON'S TRANSLATION.

"To crown the bowl," says Mr. Davidson: "sometimes signifies no more than to fill the cup to the brim, but here it is to be taken literally for adorning the bowl with flowers, according to the ancient custom. Otherwise, *implevitque mero*, would be mere tautology." Horace repeatedly speaks of crowning the bowl with Roses.

It is said that the Turks cannot endure to see a rose-leaf fall to the ground, because, says Gerarde, "some of them have dreamed that the first Rose sprang from the blood of Venus."

It may, perhaps, be worth while to quote Gerarde's translation of a passage from Anacreon, rather for its curiosity than beauty:

"The rose is the honor and beauty of flowers,
The rose is the care and the love of the spring,
The rose is the pleasure of th' heavenly powers:
The boy of fair Venus, Cythera's darling,
Doth wrap his head round with garlands of rose,
When to the dances of the Graces he goes."

This is scarcely to be recognised for the same passage given a few pages back, in the translation of one of our living poets.

Roses, when they are associated with a moral meaning, are generally identified with *mere* pleasure; but some writers, with a juster sentiment, have made them emblems of the most refined virtue. In the Orlando Innamorato, the famous Orlando puts Roses in his helmet, which guard his ears against a syren; and in Lucian, a man who has been transformed into an ass recovers his shape upon eating some Roses.*

Many species of the Rose preserve their sweet perfume

* Orlando Innamorato, Canto 33, Stanza 33. and Francklin's Lucian, vol. iii. page 236.

even after death; as the poet observes in the following passage:

> "And first of all, the rose; because its breath
> Is rich beyond the rest; and when it dies,
> It doth bequeath a charm to sweeten death."
>
> BARRY CORNWALL'S FLOOD OF THESSALY, page 2.

But nothing has yet been said to prove the assertion that poets forge chains of Roses; and were this to be omitted, many persons, considering their apparent fragility, might doubt the fact: to avoid so unpleasant a catastrophe, Tasso shall appear and speak for himself.

> "Di ligustri, di gigli, e delle rose
> Le quai fiorian per quelle piagge amene,
> Con nuov' arte congiunte indi compose
> Lente, ma tenacissime catene:
> Queste al collo, alle braccia, a i piè gli pose."
>
> TASSO, Canto 14.

> "Of *privet*, lilies, and of roses sweet,
> Which proudly flow'red through that wanton plain,
> All platted fast, well knit, and joined meet,
> She framed a soft, but surely holding chain,
> Wherewith she bound his neck, his hands, and feet."
>
> FAIRFAX'S TRANSLATION.

Fairfax translates *ligustri* woodbines: but when a foreign witness is brought into court, as Tasso is upon this occasion, it is but common justice to see that he is correctly interpreted. Suppose it had pleased the English poet to change Roses into turnips, what would have become of our cause?

We must indulge in one more quotation:

> "Ye lilies, and ye shrubs of snowy hue,
> Jasmin as ivory pure,
> Ye spotless graces of the shining field,
> And thou, most lovely rose,
> Of tint most delicate,
> Fair consort of the morn;
> Delighted to imbibe
> The genial dew of heaven,
> Rich vegetation's vermeil-tinctured gem;

April's enchanting herald,
Thou flower supremely blest,
And queen of all the flowers,
Thou formest around my locks
A garland of such fragrance,
That up to Heaven itself
Thy balmy sweets ascend."

ANDREIN'S ADAM.

Our delicate Eglantine has been scarcely less honoured by the poets than the more luxuriant Roses. It is usually coupled with the Woodbine, as the Lily with the Rose, the Myrtle with the Bay, or Beaumont with Fletcher. Shenstone, in describing the delights of a country walk after long confinement in sickness, makes particular mention of this fragrant pair:

" Come gentle air! and while the thickets bloom,
Convey the jasmine's breath divine;
Convey the woodbine's rich perfume,
Nor spare the sweet-leafed eglantine."

" Yonder is a girl who lingers
Where wild honeysuckle grows,
Mingling with the briar-rose;
And with eager outstretched fingers,
Tip-toe standing, vainly tries
To reach the hedge-enveloped prize."

H. SMITH.

" Wound in the hedge-row's oaken boughs
The woodbine's tassels float in air,
And, blushing, the uncultured rose
Hangs high her beauteous blossoms there."

MRS. C. SMITH.

The two latter passages equally apply to the Common Wild Rose; which can boast the praise of Chaucer:

" As swete as is the bramble floure
That bereth the red hepe."

Chaucer, in the Flower and the Leaf, describes a pleasant arbour formed by Sycamore and Eglantine:

> " And I, that all these plesaunt sightis se,
> Thought suddainly I felt so swete an air
> Of the eglenterè, that certainly
> There is no hert (I deme) in such despair,
> Ne yet with thoughtis forward and contraire,
> So overlaid, but it should sone have bote
> If it had onis felt this savour sote."

Keats alludes more than once to the sweet perfume of the Eglantine, when moist with rain or dew:

> " Its sides I'll plant with dew-sweet eglantine
> And honeysuckles full of clear bee-wine."
>
> ENDYMION, p. 193.

> ————————" rain-scented eglantine
> Gave temperate sweets to that well-wooing sun."
>
> ENDYMION, p. 8.

ROSEMARY.

ROSMARINUS.

LABIATÆ. DIANDRIA MONOGYNIA.

The botanical name of this plant is compounded of two Latin words, signifying Sea-dew; and indeed Rosemary thrives best by the sea.—*French,* romarin; encensier [incense-wort].—*Italian,* rosmarino; ramerino; ramarino.

ROSEMARY is common in the South of Europe, Barbary, and the Levant, and in the open ground will bear the winter in this climate; but, when in pots, it is necessary to afford it the protection of a roof during the winter season.

It has been held in high esteem as a " comforter of the brain," and a strengthener of the memory; and on the latter account is an emblem of fidelity in lovers. Formerly it was worn at weddings, and at funerals also: in some parts of England, Mr. Martyn says, " that in his time it was still customary to distribute it among the company at a funeral, who frequently threw sprigs of it into the grave." It was also planted near tombs, like Mallow and the Asphodel.

Spirit of wine, distilled from Rosemary, produces the true Hungary-water; but this is more generally made by merely dissolving the oils of rosemary and of lavender in spirit of wine. By many persons Rosemary is used as tea, for headaches and nervous disorders.

Slips or cuttings taken in the spring, just before they shoot, and planted in a pot of light fresh earth, will soon take root. When accidentally rooted in a wall or crevice of a building, it will thrive, and endure the greatest cold of our winters, however exposed to the wind.

Mr. T. Moore alludes to its character as a mourner in the following passage:

> ————————" the humble rosemary,
> Whose sweets so thanklessly are shed
> To scent the desert* and the dead."

Shenstone expresses great indignation at the little respect shown to the Rosemary in modern times:

> " And here trim rosmarin, that whilom crowned
> The daintiest garden of the proudest peer;
> Ere driven from its envied site, it found
> A sacred shelter for its branches here;
> Where edged with gold its glittering skirts appear.
> Oh wassel days! O customs meet and well!
> Ere this was banished from its lofty sphere:
> Simplicity then sought this humble cell,
> Nor ever would she more with thane and lordling dwell."
>
> School Mistress.

Shakspeare and others of our old poets repeatedly speak of Rosemary as an emblem of remembrance; and as being

* In the Great Desert are found many stalks of lavender and rosemary.—*Asiat. Res.*

worn at weddings, to signify the fidelity of the lovers. Thus Ophelia says:

"There's rosemary for you, that's for remembrance; pray you love, remember."

Again, Perdita, in the Winter's Tale:

"For you there's rosemary and rue; these keep
Seeming and savour all the winter long:
Grace and remembrance be with you both!"

Rue is the herb of grace, commonly so called in the dictionaries of Shakspeare's time.

The following passage occurs in Drayton's Pastorals:

"He from his lass him lavender hath sent,
Showing her love, and doth requital crave;
Him rosemary his sweetheart, whose intent
Is that he her should in remembrance have."

——————"will I be wed this morning,
Thou shalt not be there, nor once be graced with
A piece of rosemary."

Ram Alley, or Merry Tricks.

"I meet few but are stuck with rosemary: every one asked me who was to be married."

Noble Spanish Soldier.

In the notes to Steevens's edition of Shakspeare, many passages of this kind are quoted.

S A G E.

SALVIA.

LABIATÆ. DIANDRIA MONOGYNIA.

Salvia, from *salvere* to heal; on account of the healing qualities of these plants. Sage, from the French name, la sauge. Many of the species are also called Clary, or Clear-eye; because the seeds, powdered and mixed with honey, were supposed to clear the sight.—*French,* la sauge: at Montpellier, saoubie.—*Italian,* salvia.

MANY of the Sages are cultivated for ornament. The following are some of the handsomest:

1. The Apple-bearing.
2. The Two-coloured.
3. The Indian.
4. The Nubian.
5. The Mexican.
6. The Fulgid.
7. The Shining-leaved.
8. The Scarlet-flowered.
9. The Gold-flowered.
10. Salvia Involucrata.

The first of these has blue flowers: on the branches protuberances as large as apples are produced frequently by the puncture of an insect: these are also formed on the Common Sage in the island of Crete, where they are carried to market under the names of Sage-apples.

The second is a native of Barbary; a handsome plant, with blue and white flowers. The third, which Mr. Curtis terms a magnificent plant, has also blue and white flowers; blowing from May to July. The Nubian and the Mexican have blue flowers, blowing also from May to July. The sixth, a native of Mexico; the seventh, a Peruvian; and the eighth, from East Florida, have beautiful scarlet flowers, blowing most part of the summer. The ninth, a native of the Cape, has silver leaves and golden flowers; which are very large, and blow from May to November.

The tenth, a native of Mexico, produces an abundance of rose-coloured blossoms.

The Indian species should be in a very poor soil; so also should the seventh. The roots of the Indian Sage may be parted in spring or autumn, and both these must be sparingly watered.

The other kind may be increased by cuttings planted in spring or summer, and covered with a glass, which should be shaded from the mid-day sun. When they have taken root, they should be carefully removed into a pot of fresh earth, (a loamy soil is the best,) and again covered with the glass; as they take firmer root, the glass may be raised on one side to admit the air, and gradually withdrawn. They must all be housed in October. The Mexican kinds, in particular, must be guarded from damp. The earth should be kept moderately moist,—barely so in the winter months.

SAXIFRAGE.

SAXIFRAGA.

SAXIFRAGEÆ. DECANDRIA DIGYNIA.

French, la saxifrage.—*Italian,* sassifragia; sassifraga.

Of the Pyramidal Saxifrage there are many varieties: the flowers are mostly white, dotted with red; and when the roots are strong, they will produce large and handsome pyramids of them, blowing in June, and making a showy appearance. If placed in the shade, and screened from wind and rain, they will longer preserve their beauty. This species is from the Alps and Pyrenees.

The Saffron-coloured Saxifrage grows on the mountains in Switzerland, Carniola, and Italy. They produce plenty of offsets on the sides of the old roots, which should be planted in a fresh light earth: they should stand in the shade in the summer, and in the sun in winter. All the offsets should be removed, which will cause them to shoot a stronger flowering stem. These being planted in small pots, and removed the next year into larger, will then be in a condition to flower: the old roots perish after flowering. The Saffron-coloured must be sheltered from frost.

The Thick-leaved Saxifrage bears purple flowers: the stem changes every year into root; losing its leaves in the winter after flowering, turning to the ground, and changing black, where it puts out fibres for the succeeding plant. The foliage of this species is remarkably handsome: it is a native of Siberia, and flowers in April and May. A variety of this, called the Heart-thick-leaved, produces larger flowers. This species prefers a rich moist soil. The roots may be parted in spring or autumn: if the winds are cold when it is in flower, it should be removed into the house.

The species called None-so-pretty is a native of Ireland, and, as it is said, of England: but it was not known to be indigenous till long after it had been cultivated in our gardens, where it was much admired for its flowers, for which it obtained its familiar name. It has also been called London-pride, from thriving well in the smoke of London, which some of the Alpine Saxifrages will not do. The flowers are white or flesh-coloured, dotted with yellow and dark red: they blow in June and July.

The Purple-flowered Saxifrage grows naturally upon rocks, which, with its numerous trailing branches, it clothes with a rich tapestry, in the months of April and May. In gardens it blows in February or March. The flowers are large and handsome, and the more exposed the situation, the greater number they produce. There is a variety of this kind called the Biflora, or Two-flowered; which, as this name implies, produces only two flowers on one stem, but those are of a beautiful rose-red. Towards the end of March divide a plant, which has filled the pot the year before, into many small pieces, observing that each piece has two or three fibres: plant half a dozen of these in the middle of a small pot, filled with bog-earth and loam, equal parts of each: water it, and place it in the shade for a week; then expose it to the morning sun, water it once a day in dry weather, and in the spring the pot will be covered with a profusion of bloom.

This should be treated every year in the same manner. It is very hardy, and disdains all tender treatment.

The Round-leaved Saxifrage is a native of Austria, Switzerland, Piedmont, &c. "We know of no species," says Mr. Curtis, "belonging to this beautiful genus, whose flowers, in point of prettiness, can vie with these."

The roots should be parted in autumn: they require a stiff loamy soil and a shady situation, and must be kept moist.

White Saxifrage produces its flowers in April and May: the Double-flowered variety is very commonly planted in pots, to adorn halls, windows, &c. in the spring. It produces plenty of offsets; and in July, after the leaves have decayed, these should be taken off, and planted in fresh unmanured earth. Till autumn it should be placed in the shade, then removed into the sun, where it should remain till the end of winter.

SCABIOUS.

SCABIOSA.

DIPSACEÆ. TETRANDRIA MONOGYNIA.

French, la scabieuse des Indes; regardez moi [look at me]; fleur de veuve [widow's flower].—*Italian,* scabbiosa gentile; fior della vedova.

Indian or Sweet-scabious is chiefly valuable for its exceeding sweetness; yet its colours are often extremely rich. It is sometimes of a pale purple, sometimes so dark as to be almost black, but its finest hue is a dark Mulberry red.

If Scabious is sown in March, it blows in the autumn; but it will produce stronger flowers if sown in May, placed in the shade, and, when come up, removed into fresh earth; if well watered and shaded till it is again rooted, and always kept moderately moist, it will flower in the beginning of the summer; and by this management may be preserved in beauty from June till September.

Though this is frequently called the Indian Scabious, botanists are uncertain of its native country; hesitating between Spain, Italy, and India.

Many persons transplant Scabious a second time at Michaelmas, for it is one of those plants which are thought to be benefited by removal.

SCARLET-BEAN.

PHASEOLUS MULTIFLORUS.

LEGUMINOSÆ. DIADELPHIA DECANDRIA.

Called also Scarlet-runners.

THE species of bean, commonly called the Scarlet-runner, will thrive well in a deep pit or box; and is well worthy of attention for the beauty of its blossoms. It will clothe whole walls or fences, for a time, with a luxuriant green and red tapestry. If sown in pots, one seed will suffice for each; but the better way is to have a box of some length, placed against the wall of a court, area, &c., and there to sow the seeds, about six inches apart and an inch deep: this should be done towards the end of April, or early in May. Sprinkle the earth with water on sowing the seed; after which, be sparing of it till the plants begin to shoot: they should then, in dry weather, be watered three times a week. When the plants have risen six or eight inches, sticks should be placed to support them; unless they are against a wall, which may serve for that purpose, and they will quickly spread over it in luxuriant beauty. Before Miller's time, it was cultivated less for its fruit than for the beauty and durability of its blossoms, which the ladies put into their nosegays and garlands. He brought it into general use for the table; and, because it has been found so useful, people seem to think it can no longer be ornamental, which is surely a vulgar mistake.

SEA-LAVENDER.

STATICE.

PLUMBAGINEÆ. PENTANDRIA PENTAGYNIA.

The Scolloped-leaved Sea-lavender has a yellow flower with a handsome blue calyx, handsomer than the flower itself. It retains its beauty when dry, and is so preserved with other flowers in winter.

The Rough-leaved, a native of Barbary and the South of Europe, has pale blue flowers. The Plaintain-leaved, a Russian, has white flowers: these three kinds are biennial.

The Triangular-stalked, from the Canary islands; the Narrow-leaved-shrubby, from Siberia; and the Broad-leaved-shrubby, from Sicily, are perennials: these may be increased by cuttings planted in July.

They should be kept moderately moist; and, with the exception of the third, must be sheltered in the winter, admitting fresh air in mild weather.

Thrift, which was the predecessor of Box as an edging for flower-borders, is of this genus: the bright scarlet variety is very pretty. It is named Thrift from its readiness to thrive in any soil, situation, climate, air, fog, or smoke. It is also called Mountain-pink, Sea-pink, Lady's-cushion, and Sea-gilliflower: it may be increased by parting the roots in autumn. The French call it, *gazon d'Espagne* [Spanish turf]; *gazon d' Olympe* [turf of Olympus]; *oeillet de Paris* [Parisian pink]; *herbe à sept tiges* [herb with seven stalks]: the Italians, *statice; pianta da sette fusti* [plant with seven stalks].

SEDUM.

CRASSULACEÆ. DECANDRIA PENTAGYNIA.

The Sedums include the Stonecrops and Orpines: many persons are very curious in these plants, which are adapted principally for veiling unsightly walls, enriching cottage-roofs, or wall-tops, or dropping from the eaves. Some of them are very splendid.

The Orpine Stonecrop is also called Live-long, because a branch of it hung up will long retain its verdure; but this is common to most very succulent plants, which will feed for a long time on the moisture they have previously imbibed. It is common in Europe, Japan, and Siberia; and is called in France, *la reprise; grassette; feve epaisse* [thick bean]; *Joubarbe des vignes* [vine Jupiter's beard]; *feuille gras* [thick leaf]; *herbe magique* [magical herb]: and in Italy, *sopravivolo; telefio; fava grassa; favogello; pignuola.*

The Thick-leaved species with white flowers makes a beautiful appearance all the year round, and spreads fast:

"Cool violets, and orpine growing still."

The Sedum Reflexum, Trip-madam, or Yellow Stonecrop—called by the French, *trippe madame*—is common all over England, on walls, and thatched roofs, where it spreads a continual vegetable sunshine: it flowers in July. Haller says this kind is eaten in salads.

The Biting Stonecrop, Pepper-crop, or Wall-pepper, so called from its pungency, is also very common in England. It is either planted on walls, or in pots, placed in a lofty situation, from whence it hangs over the sides of the pots, and grows to a considerable length. It is called by the French, *la vermiculaire brulante* [burning wormwood];

pain d'oiseau [bird bread]; in the village dialect, *perratin:* and by the Italians, *erbi pignuola; pinocchiella.*

The Orpines may be readily increased by cuttings in the summer months. The flowers are generally white or purple. The Evergreen kind, with purple flowers, spreads very fast. They require a dry soil, and prefer the shade.

Any of the perennial Stone-crops planted in a little soft mud or earth, in the manner of the House-leeks, on a wall or roof, will spread its roots, and cover the whole place in a short time. Of the annual kinds, the seeds sprinkled over the place where they are designed to grow will easily root. When planted in pots, they should always be placed high from the ground. They are very succulent plants, and will not require watering.

SHADDOCK TREE.

CITRUS DECUMANA.

AURANTIACEÆ. POLYADELPHIA POLYANDRIA.

French, la pampelmouse.—*Italian,* pamplemusa; these names are from the Dutch. The Italians also call it pompa di genova.

This plant may be treated in the same manner as the Orange; under which head, directions are given at full length.

The Shaddock is a native of India; it was taken to the West-Indies, in an East-India Ship, by Captain Shaddock from whom it was named. There are many varieties; one, bearing a fruit five inches in diameter, very sweet, is called in China Sweet-Ball. In Japan, the Shaddock is said to be as large as the head of a child.

The trees imported from Italy, and sold at the Italian

warehouses in London for Orange-trees, without any peculiar name, are for the most part either Shaddock or Citron-trees, as these sorts make stronger shoots and more showy plants than the true Orange.

SNOW-DROP.

GALANTHUS.

NARCISSEÆ. HEXANDRIA MONOGYNIA.

Galanthus is of Greek origin, and signifies Milk-Flower: the flower being very white. Its name of snow-drop expresses the same thing, and is, at the same time, applicable to the time of its appearance, often when snow is on the ground. In mild seasons it will blow in January, but it usually appears in February, on which account it has also been named Fair-Maid of February.—*French,* la galantine;—*Italian,* galanto.

The Snowdrop is a native of Switzerland, Austria, Silesia, and England, in meadows, and orchards; but doubts are entertained whether it is really indigenous, or whether it is a relic of cultivation. Every third year, the roots should be taken up in June, when the leaves have decayed, and kept in a dry place till August; they should then be replanted; and the best way to make them look well, is to plant twenty or more together in a clump, which has a very pretty effect when they blow. They should not however be less than an inch and a half apart, and should be set two inches deep.

There is a flower called the Leucojum, or great Snow-drop, very similar to this, but twice its size. Of this there are three kinds, commonly called the Spring, the Summer, and the Autumnal Snow-drop. Some to distinguish them better, being of a different genus, have named them Snowflakes; others Bulbous White-Violets; but the kind which one calls the early-flowering Bulbous White-Violet, in reference to a kind flowering later, another calls the late-flowering, in reference to one blowing earlier, which occasions infinite confusion. These are what the French call, *perce neige* [snow piercer]; the Spring kind being also called *violette de Fevrier* [February Violet]; *violier bulbeux,* [bulbous stock]; *campane blanche, cloche blanche* [both signifying white bell]; *baguenadier d'hivèr.*

These flowers are very pretty and delicate, and look well like the common Snow-drop whèn planted several together, but it must not be close; for they require a distance of five inches from each other, and must be set four or five inches deep. Thus they require more room than will often be afforded them, except in the open ground; and, after all, they are deficient in one of the greatest charms of the true Snow-drop—the coming in a wintry season, when few others visit us. We look upon the Snow-drop as a friend in adversity; sure to appear when most needed.

The Snow-drop is the earliest blower of all our wild flowers, and will even show her head above the snow, as if to prove her rivalry in whiteness.

"As Flora's breath, by some transforming power,
Had changed an icicle into a flower."

Mrs. Barbauld.

"Like pendent flakes of vegetating snow,
The early herald of the infant year,
Ere yet the adventurous crocus dares to blow
Beneath the orchard boughs thy buds appear.

"While still the cold north-east ungenial lowers,
And scarce the hazle in the leafless copse
Or sallows show their downy powdered flowers,
The grass is spangled with thy silver drops.

"Yet when those pallid blossoms shall give place
To countless tribes of richer hue, and scent,

Summer's gay blooms, and Autumn's yellow race,
I shall thy pale inodorous bells lament.

" So journeying onward in life's varying track,
Even while warm youth its bright illusion lends,
Fond memory often with regret looks back
To childhood's pleasures, and to infant friends."
MRS. C. SMITH.

SOUTHERNWOOD.

ARTEMISIA ABROTANUM.

CORYMBIFERÆ. SYNGENESIA POLYGAMIA SUPERFLUA.

From Artemisia, the wife of Mausolus, King of Caria; called also Old Man.—*French*, l' auronc-des-jardins; la citronelle; la garde-robe, from its use in preventing moths from getting into wardrobes and clothes-presses.—*Italian*, Abrotano, abruotino, abruotina.

SOUTHERNWOOD is well known as an aromatic shrub, growing three or four feet high. It is a native of many parts of Europe and Asia, where it produces an abundance of small yellow flowers; but the flowers seldom open in this country.

It may be increased by slips planted in April, and well watered: they must remain in the shade till rooted. This plant is often esteemed by old persons for its aromatic scent; but is not now a very *fashionable* plant. It was formerly a common garden plant in London, as it will live even in the densest parts. It is used in medicine, and its branches will dye wool yellow.

The Artemisia is included among the flowers of poetical origin in Mr. Smith's Poem of Amarynthus:

" *That* with the yellow crown named from the queen
Who built the Mausoleum."

SPEEDWELL.

VERONICA.

RHINANTHACEÆ. DIANDRIA MONOGYNIA.

French, veronique.—*Italian,* veronica.

Most of the Veronicas are natives of cold countries, and consequently hardy: they may be increased by parting the roots in autumn; which, in pots, should be done every year. The annual kinds may be sown in autumn.

The Cross-leaved species requires shelter from frost; it is increased by cuttings made in any of the summer months. These plants prefer the shade, and must be kept moist.

The flowers are flesh-coloured, blue, or white. The Blue Rock Speedwell is a beautiful little plant, and is a native of Switzerland, Austria, Denmark, Norway, and Scotland. It is by some familiarly called Forget-me-not; a name given also to the Ground Pine, a species of Germander: but the true Forget-me-not is the water mouse-ear, the myosotis palustris of the botanists.

The Germander Speedwell is a native of Europe and Japan. "Few of our wild flowers," says Mr. Martyn, "can vie in elegance and brilliancy with this; and many plants with far less beauty are cultivated in our gardens. In May and June every hedge-bottom and grassy bank is adorned with it. At night, or under the influence of moisture, the corolla closes, but in dry bright weather appears fully expanded; and though each flower is short lived, there is a copious succession."

Dr. Withering says the leaves are an excellent substitute for tea. The Common-Speedwell has been much recommended for this purpose, especially in Germany and Sweden; and the French still call it the *Thè de l'Europe.*

SPIRÆA.

ROSACEÆ. ICOSANDRIA PENTAGYNIA.

The name Spiræa signifies a rope, these shrubs being flexible like ropes, and also because many parts of the stem, and the fruits of some of the species, are twisted. It is also called Bridewort.

This is a beautiful genus; most of the species are handsome flowering shrubs: the Willow-leaved, commonly called Spiræa Frutex, grows to a height of from three to six feet, according to the soil; the blossoms are handsome, and of a rose-red; blowing in June and July. In moist seasons, the young shoots from the root will frequently flower in autumn. It is a native of Siberia.

The Scarlet Spiræa is a native of Pensylvania; the blossoms are of a beautiful red-colour, blowing in August and September.

It is not determined whether the Hypericum-leaved Spiræa be a native of Italy, or of North America; it is called Hypericum Frutex, and Italian May. In Italy, there are hedges of it, bearing a profusion of blossoms. It flowers in May and June. Its height is five or six feet.

The Germander-leaved kind also makes beautiful hedges. The Kamschadales use the leaves of this as tea, and make tobacco-pipes of the straight shoots: it flowers in June.

The Three-lobed-leaved kind is a Siberian, it grows about two feet high, bears white flowers, and is a very elegant plant.

The Currant-leaved Spiræa, familiarly called the Virginian Guelder Rose, grows nine or ten feet high: the blossoms are white, spotted with pale-red.

The Spiræa Filipendula, or Dropwort, is an herbaceous plant; so called from the manner in which its tuberous roots hang together by threads. The flowers are cream-coloured, often tipt with red, opening in July. It grows about a foot and a half high, and sometimes produces double flowers.

The Spiræa Ulmaria, Meadow Sweet, or Queen of the Meadows—called in French, *la reine des prés; l'ormiére, vignette* [little vine]; *petite barbe de chèvre* [little goats beard]: and in Italian, *ulmaria; regina dei prati*—is likewise an herbaceous plant; it abounds in moist meadows, perfuming the air with the Hawthorn-like scent of its abundant white blossoms, throughout June, July, and August. It grows three or four feet high. There is a variety with double flowers.

The most elegant kind is the Three-leaved Spiræa, but that is very difficult to preserve: it should be in a bog, or peat earth, and in a shady situation.

Of the shrubby Spiræas, the dead wood and the irregular branches should be pruned off every year: the suckers should likewise be removed, or they will starve the old plant. They should be new-potted, and have fresh earth given them every spring.

Of the herbaceous sorts the roots may be parted in autumn. They must all be kept moderately moist. Being generally natives of cold countries, they do not fear the cold.

Clare mentions the Meadow Sweet, in speaking of the effect of the noon-day sun upon flowers in the open country:

> "Oh! to see how flowers are took,
> How it grieves me when I look:
> Ragged-robins, once so pink,
> Now are turned as black as ink,
> And the leaves, being scorched so much,
> Even crumble at the touch;
> Drowking lies the meadow-sweet,
> Flopping down beneath one's feet:
> While to all the flowers that blow,
> If in open air they grow,

Th' injurious deed alike is done
By the hot relentless sun.
E'en the dew is parched up
From the teasel's jointed cup."

STAR OF BETHLEHEM.

ORNITHOGALUM.

ASPHODELEÆ. HEXANDRIA MONOGYNIA.

The botanical name of this genus is from two Greek words, which signify *bird* and *milk*; which Mr. Martyn supposes to be intended to express the whiteness of the flowers or roots of some of the species, like the feathers or beaks of some birds. But this is surely a long way to fetch a name, or its explanation.—*French*, l'ornithogale; churles.——*Italian*, ornitogalo; latte d' uccello.

THE Snowy Star of Bethlehem, the Spear-leaved, the Long-spiked, the Cape, the Grass-leaved, and the Golden, are all from the Cape, and too tender to thrive in the open air. In the beginning of July, when the leaves and stalks decay, the roots may be taken up, and laid in a dry place till the end of August, when they must be planted again. They may be increased by offsets.

The Pyrenean, Close-spiked, Broad-leaved and Pyramidal kinds, are hardy bulbs: they may be increased by offsets, which they produce in great plenty. They should be transplanted in July or August, but not oftener than every second year. They should have a light sandy soil.

All the kinds must be kept moderately moist.

STAR LILY.

AMARYLLIS.

AMARYLLIDEÆ. HEXANDRIA MONOGYNIA.

French, amarillis.—*Italian*, giglio narciso [narcissus lily].

THE name of Amaryllis is supposed to be derived from

a Greek word signifying splendour; "and is given," says Mr. Martyn, "with great propriety, to this splendid genus."

The Yellow Amaryllis, or Autumnal Narcissus, is a native of the South of France, Spain, Italy, and Thrace. The flowers seldom rise above four inches high; and somewhat resemble the Yellow Crocus. Like that, too, its leaves grow all the winter, after the flowers are past. It flowers in September, is very hardy, and increases fast by offsets. They may be transplanted any time from May to the end of July, but not later.

This plant prefers a light dry soil, and an open situation. It must not be under the dripping of trees. In mild seasons, there will often be, from the same root, a succession of flowers from September to the middle of November. It should be kept moderately moist.

The Atamasco Lily is a native of Virginia and Carolina, where it grows plentifully in the fields and woods, and makes a beautiful show. At their first appearance the flowers are of a fine carnation colour outside, but they fade almost to white: they blow from May to July or August.

It may be increased by offsets: the bulbs should be removed every second year, and, if they begin to shoot while out of the earth, should be planted immediately. It should be kept moderately moist.

The Jacobœa Lily—in French, *le lys de St. Jaques* [St. James's lily]; *la croix de St. Jaques* [St. James's cross]; *la belle amarillis:* and in Italian, *giglio narciso giacobeo*—produces its flowers two or three times in the year, not at any regular season. It furnishes plenty of offsets, which should be taken off every year: the best time is in August, that they may take good root before winter. In removing the roots, great care should be taken not to break off their fibres. This flower may stand abroad in the summer, but in the winter should be lodged in an inhabited room. It must be kept moist.

This Lily is a native of South America: the flowers are large, of a deep red, and bend gracefully on one side of the stalk. Parkinson calls it the Indian Daffodil.

The Belladonna Lily—called by the French, *lis de Mexique* [Mexico lily]; *la belle dame:* and by the Italians, *narciso bella donna* [fine lady narcissus]—is a native of the West Indies, and grows on shady hills, and by the margins of streams. It is of a pale purple colour, inclining to white towards the centre. It was first brought to England from Portugal, and is very common in the Italian gardens, particularly in the neighbourhood of Florence, where it is sold in the markets under the name of Narcissus-belladonna. This Lily is very fragrant. It flowers about the end of September or the beginning of October, and, if the weather be favourable, will continue in bloom a month, or more.

In June the leaves decay, and the root should be transplanted soon after; for, if it remains till July, it will send forth new fibres; and removal then would injure it. It should remain in the house in the winter, and be kept moderately moist.

The Superb, or Riband Amaryllis, is supposed to be a native of the Cape: the flowers are very beautiful; a white ground, striped with red. Unless hastened by artificial heat, they open in April or May. As this bulb rarely produces offsets, it should be procured in the pot, and treated as the last.

The Long-leaved Lily, or Amaryllis, is a native of the Cape of Good Hope. The flower-stem is seldom more than four inches high, but bears a profusion of purple flowers, opening in December. It may be treated as the Jacobœa Lily.

The Guernsey Lily—called in France, *le lis de Japon*—which has been removed by some botanists from the genus Amaryllis, and called Nerine, is extremely handsome: it is a native of Japan, but has long been naturalized at Guernsey, from which place it is named. There are from eight to twelve flowers on one plant; the circumference of each flower about seven inches. When in full beauty, it has the appearance of a fine gold tissue wrought on a rose-coloured ground; and when it begins to fade, it is pink. If beheld in full sunshine, it seems studded with diamonds; but by candle-light, looks rather as if it were spangled with fine gold-dust. When the flower begins to wither, the petals assume a deep crimson colour. The flowers begin to appear towards the end of August, and the head is usually three weeks gradually expanding. This plant is said to have been taken to Guernsey by a vessel wrecked there on its return from Japan. There, and at Jersey, it thrives as well as in its native country; and, from both those islands, the roots are annually dispersed over Europe.

These roots, or rather bulbs, are generally brought over in June or July: they should then be planted in pots of light earth, and refreshed with water two or three times a week, but very gently. Too much wet, especially before they come up, would rot the bulbs.

About the middle of September, such of the bulbs as are strong enough to flower will begin to show the bud of the flower-stem, which is commonly of a red colour: they should then be placed where they may have the benefit of the sun, and be defended from strong winds; but by no means must they be placed close to a wall, or under glasses, which would draw them up weak, and render them less beautiful. If the weather be dry, they should still be refreshed with water every second, or if very hot, every day; but if there be much rain, they must be sheltered from it.

When the flowers begin to open, they should be placed under cover to preserve them from rain; but must be allowed plenty of fresh air, or the colours will lose their brilliancy, and soon decay. If rightly managed, they will continue in beauty a full month; and, though they afford no perfume, their beauty alone entitles them to a first rank among the children of Flora.

After the flowers have decayed, the leaves will continue growing all the winter; they must be defended from frost, but should have as much free air as possible in mild weather: when it is both mild and dry, they may stand abroad in the middle of the day. The roots should not be removed oftener than every fourth year, towards the end of June, or early in July; they should then be replanted in fresh earth, and the offsets planted in separate pots. These young plants will produce flowers the third year.

The bulbs of this Lily do not flower every succeeding year, as most bulbs do; but if they contain two buds in the centre, as is often the case, they will flower twice in three years; after which the same root will not flower again for several years, but only the offsets from it.

STOCK.

MATTHIOLA.

CRUCIFERÆ. TETRADYNAMIA SILIQUOSA.

Stock-gilliflower.—*French,* giroflée; violier.—*Italian,* viola; leucoio.

The Virginia Stock is improperly so called: its native place is on the coast of the Mediterranean, and it is called in French, *giroflée de Mahon:* it is an annual plant, too well known to need description. The seeds of this may be sown at two or three several times, to obtain a longer succession of flowers; in autumn, and in March, April, or May. It does not rise above six inches high; but, as it branches, three or four seeds will be enough for a middle-sized pot.

The kind commonly called the Queen's Stock-gilliflower —in French, *giroflée des jardins* [Garden Stock]—varies in colour from a pale to a deep red, and is sometimes variegated; but the bright red is most esteemed. As this branches very much, one seed only must be sown in a pot: this should be done in May; water should be given every evening; and, during the heat of the day, the pots should be shaded, to prevent the earth from drying too fast. They must be protected from frost during the winter, either by removing them into the house, or covering them with oak-leaves. The poorer the soil in which they are planted the better they will bear the cold. The following May they will flower, which they often continue to do all the summer, and probably many of the flowers will come out double. In autumn, after they have blown, they usually perish; but when they are in a very poor soil, or are growing among rubbish, they will often last two or three years.

The Brompton—in French, *giroflée à tige*—and the White Stock, are varieties of this kind; the latter will sometimes live three or four years. This species is a native of the coast of Spain, Greece, Italy, Candia, and the isles adjacent.

The Stock-gilliflower has been long established in the English gardens, and is indeed a native of the cliffs by the sea-side. The old English name of Gilliflower, which is now almost lost in the prefix, Stock, is corrupted from the French *giroflier.* Chaucer writes it Gylofre; but, by associating it with the nutmeg and other spices, appears to mean the Clove-tree, which is, in fact, the proper signification of that word.

Turner calls it Gelover and Gelyfloure; Gerarde and Parkinson, Gilloflower. Thus, having wandered from its original orthography, it was corrupted into July-flower. Pinks and Carnations have also the title of Gilliflower from smelling like the clove, for which the French name is *girofle*. For distinction, therefore, they were called Clove-gilliflowers, and these Stock-gilliflowers. Gerarde adds the names Castle-gilliflower, and Guernsey-violet.

The Annual, or Ten-weeks' Stock—French, *le quarantain; le violet d'été* [summer violet]: Italian, *leucoio estivo* [summer stock]—grows about two feet high: there are many varieties, white, red, purple, and striped; and double and single varieties of each of these colours. It grows naturally on the coast in the South of Europe. By means of a hot-bed they may be raised earlier, but without that help the best season for sowing them is in March and April, and indeed in May also: if they are taken in when the weather becomes severe, they will continue to flower; those planted in May will last to the very end of winter, in the house. A middle-sized pot will contain three or four.

The Broad-leaved Shrubby-stock is a native of the island of Madeira; it blossoms from March to May: when the flowers first open, they are white, sometimes inclining to yellow; in a few days they become purple; hence this species has been termed *mutabilis*, or changeable. This is of quick growth, and may be increased by cuttings, taken as soon as the plant has done flowering: they should be housed in the winter.

Some persons increase the Queen's-stock in the same manner, planting the cuttings in March or April in pots three or four inches wide; in the middle of May they remove them into pots of five or six inches diameter, and in July or August into full-sized ones, that is, eight or ten inches; but though these cuttings will generally root, they do not make such handsome plants as those raised from seed: it is not, therefore, worth while to practise this method unless to preserve some fine double flowers. These flowers love the sun; but care must be taken to supply in the evening the moisture which has been exhausted during the day. It will be observed, too, as an invariable rule, always to place a plant in the shade when newly potted, and to let it remain there till rooted.

There are other species of Stock, but these are the most desirable. There is a Cheiranthus, called the C. Quadrangulus, a native of Siberia, which was introduced into the Paris garden by Jean Jacques Rousseau. The flowers are sulphur-coloured and sweet. It is propagated by seeds, and thrives in the open air, but does not last many years.

STRAMONIUM.

DATURA.

SOLANEÆ. PENTANDRIA MONOGYNIA.

Called also Thorn-apple.—*French,* stramonie; la pomme epineuse; herbe aux sorciers; herbe des magiciens [both signifying conjurors'-wort]; endormie [sleeper]; herbe du diable [devil's-wort]; pomme du diable [devil's-apple]; herbe a la taupe [mole-wort]; noix metelle [metel-nut] which last properly belongs to the datura metel.—*Italian,* datura; pomo spinoso [thorny-apple]; stramonio; noce metella.

Some few of the Stramoniums require the protection of a stove: the other kinds are usually raised in a hot-bed. The Purple Stramonium is the handsomest: the flowers are purple on the outside, and of a satiny white within;

and blow in July. The double-flowered varieties are the most esteemed.

STRAWBERRY-BLITE.

BLITUM.

ATRIPLICEÆ. MONANDRIA DIGYNIA.

Blitum is derived from the Greek, and signifies, *fit only to be thrown away:* it is also called Strawberry Spinach, and Berry-bearing Orach. *French,* bléte; arroche.

The name of these plants may not appear very inviting; but it is to be understood with some limitations: they bear fruit resembling the Strawberry in appearance, and all the name is intended to imply is, that the fruit is unfit to eat. Having thus explained matters, I will proceed to introduce the plants themselves, which, perhaps, may make a more favourable impression than if more expectation had been excited.

There are three or four species of the Strawberry-blite, all annuals, and easily raised from seed. They may be sown in March or April, three or four seeds in a pot of eight or nine inches diameter, of the Swedish kind; but only one, of the others. In five or six weeks the plants will come up, and in July will begin to show their berries. They should always be kept moderately moist, and must stand in the open air. As the flower-stems advance in height, they will require sticks to support them, or the weight of the berries will bear them down.

The Blitum Capitatum, or Berry-headed Strawberry-blite, bears the largest berries.

SUN-FLOWER.

HELIANTHUS.

CISTEÆ. SYNGENESIA POLYGAMIA FRUSTANEA.

French, l'hélianthe; fleur du soleil; soleil [the sun]; tournesol [sun-turner]; couronne du soleil [crown of the sun]; herbe du soleil [sun-wort].—*Italian,* girasole; fior del sol; corona del sole; girasole Indiano [Indian sun-turner]; girasole Peruano.

THE Sun-flower can scarcely be introduced here with propriety, being in general so large, even the annual kinds, as to be ill-adapted for pots. The Annual Sun-flower rises to the height of twelve or fourteen feet, and the flower sometimes exceeds a foot in diameter. It is not called Sun-flower, as some have supposed from turning to the sun, but from the resemblance of the full-blown flower to the sun itself: Gerarde remarks, that he has seen four of these flowers on the same stem, pointing to the four cardinal points. This flower is a native of Mexico and Peru, and looks as if it grew from their own gold. It flowers from June to October.

The Dwarf Annual kind, which grows from eighteen inches to three feet in height, is a little more within compass.

The Perennial Sun-flower is much esteemed for bouquets; the flowers are about eight or ten inches in diameter: there is a constant succession from July to November. It is a native of Virginia.

The Dark-red Sun-flower, and the Narrow-leaved, are of a more moderate height; the first, two or three feet, the latter, a foot and a half. Both are natives of Virginia, flowering in September and October.

The Sun-flowers are hardy plants; the perennial kinds are increased by parting the roots into small heads: this should be done in the middle of October, soon after the flowers are past, or very early in the spring, that they may be well rooted before the droughts come on. They will require watering in dry weather, particularly when in pots.

Several of the Sun-flowers are natives of Canada, where they are much admired and cultivated by the inhabitants, in gardens, for their beauty: in the United States they sow whole acres of land with them, for the purpose of preparing oil from their seeds, of which they produce an immense number. This oil is very pure, fit for salads, and for all the purposes of Florence oil*.

Thomson supports the popular notion that this flower turns ever towards the sun:

> " Who can unpitying see the flowery race,
> Shed by the morn, their new-flushed bloom resign,
> Before the parching beam? So fade the fair,
> When fevers revel through their azure veins.
> But one, the lofty follower of the sun,
> Sad when he sets, shuts up her yellow leaves,
> Drooping all night, and, when he warm returns,
> Points her enamoured bosom to his ray."

Mr. T. Moore has taken advantage of the same idea, in the words of one of his Irish Melodies:

> " As the sun-flower turns to her god when he sets
> The same look which she turned when he rose."

Clare gives a natural picture of the Sun-flower in the following description of the floral ornaments of a rustic cottage:

> " Where rustic taste at leisure trimly weaves
> The rose and straggling woodbine to the eaves,—
> And on the crowded spot that pales enclose
> The white and scarlet daisy rears in rows,—
> Training the trailing peas in bunches neat,
> Perfuming evening with a luscious sweet,—

* See Lambert's Travels in Canada, &c.

And sun-flowers planting for their gilded show,
That scale the window's lattice ere they blow,
Then, sweet to habitants within the sheds,
Peep through the diamond panes their golden heads."

VILLAGE MINSTREL, &c. vol. ii. page 80.

The size and splendour of this flower make it very conspicuous, and some have accused it of being gaudy, although constant in the one golden colour of its attire: gaudiness, however, is a quality which may be pardoned in a flower,

"Where tulip, lily, or the purple bell
Of Persian wind-flower; or farther seen
The gaudy orient sun-flower from the crowd
Uplifts its golden circle."

MATURIN'S UNIVERSE, page 55.

The Sun-flower was formerly called Marygold also, as the Marygold was termed Sun-flower. Gerarde styles it the Sun-marygold.

There is another genus producing the same kind of flowers, only smaller, usually called the Willow-leaved Sun-flower. Their botanical name is Helenium, supposing them to have sprung from the tears of Helen, the wife of Menelaus: it has not been clearly ascertained upon what occasion. Drummond speaks of this flower in his lines on the death of Prince Henry:

"Queen of the fields, whose blush makes blush the morn,
Sweet rose, a prince's death in purple mourn;
O hyacinth, for ay your Ai keep still,
Nay with more marks of woe your leaves now fill:
And you, O flower! of Helen's tears that's born,
Into those liquid pearls again now turn."

SWEET-PEA.

LATHYRUS.

LEGUMINOSÆ. DIADELPHIA DECANDRIA.

French, pois odorans; pois de senteur [both signifying scented pea]; pois de fleur [flower pea.]

The Sweet-pea has several varieties, greatly differing in colour: the common sort, which is blue and dark-purple, sometimes with a tinge of red, is a native of Sicily. The more delicate kind, white and blush, or white and deep rose-colour, sometimes with a mixture of pale blue, is a native of the Island of Ceylon, and is called the Painted-lady.

The Tangier-pea is a native of Barbary, its colours purple and red: it is an annual plant, which grows to the height of four or five feet; blossoms in June or July, and dies in autumn.

Although the Sweet-pea is now so common in this country that we seldom see a garden, however small, that cannot boast of possessing it; it is not more than a hundred years since it was numbered among our rare and curious plants, and in the time of Parkinson and Evelyn it was not known in our gardens.

This Pea blows in June, and continues in blossom till killed by the frost. It may be sown about half an inch deep, and it may be well to scatter the seeds pretty thickly; if they all grow, the weaker ones may be removed, and the stronger left. They may be sown in October, and kept in-doors till spring; or may be sown and placed abroad at once in March or April. In cold weather, the earth should be just kept moist; in hot dry summer weather, it must be watered every evening, and if necessary in the morning also. When the plants are about three inches high, sticks should be placed to support them, three or four feet in length. This plant should not be kept within doors in warm weather, or it will grow very tall and weakly, and produce few flowers.

There is a variety of this Pea entirely white; but the most beautiful is the red and white. But that I fear to confess so great a heresy, I would say this flower need not yield even to the rose. Nothing can exceed the elegance of its form; nor can there be a more delicate contrast of colour. They are justly termed Papilionaceous, for they do indeed look like butterflies turned to flowers. It is sometimes difficult to believe that the little white butterflies which reel about in the sunshine are not white violets or peas which have broken their bonds. It is equally difficult to believe that these flowers want any thing but will to fly: and we almost expect to see them start from their stalks as we look at them.

Both these fancies are authorised by the poets.

> " In their own bright Kathaian bowers
> Sparkle such rainbow butterflies,
> That they might fancy the rich flowers,
> That round them in the sun lay sighing,
> Had been by magic all set flying.
>
> LALLA ROOKH.

These butterflies, Mr. Moore tells us, are called, in the Chinese language, Flying-leaves. " Some of them," continues he, " have such shining colours, and are so variegated, that they may be called Flying-flowers ; and indeed they are always produced in the finest flower-gardens."

> " Here are sweet-peas on tip-toe for a flight,
> With wings of gentle flush, o'er delicate white,
> And taper fingers catching at all things
> To bind them all about with tiny rings."
>
> KEATS.

In his Calendar of Nature Mr. Hunt speaks of Sweet-peas, as looking like butterflies turned to flowers.

In short, it seems scarcely possible not to feel this. They seem only lingering to sip their own honey.

SYRINGA.

PHILADELPHUS.

MYRTEÆ. ICOSANDRIA MONOGYNIA.

From Ptolemy Philadelphus, King of Egypt. It is also called mock-orange, and pipe-tree.——*French,* le seringat; in Languedoc, siringea:—*Italian,* siringa.

THE Syringa is a most delicious shrub: the foliage is luxuriant, the blossoms beautiful, and abundant, white as the purest lily, and of the most fragrant scent; in a room, indeed, this perfume is too powerful, but in the air, it is remarkably agreeable. There is a variety which has no scent; and also a dwarf variety, which does not usually exceed three feet in height: the flowers are sweet, and double; but it flowers rarely, and is on that account less esteemed than otherwise it would be.

There is a species called the Myrtle-leaved Syringa, a native of New Zealand; the fresh flowering shoots of which were used as tea by Captain Cook's sailors, who found the infusion sweetly aromatic at first; in a short time, however, it became very bitter. It was considered serviceable in the sea-scurvy.

The Sweet Syringa, specifically so called, is also a native of New Zealand. It flowers in July and August; the Myrtle-leaved kind in June and July; and the Mock-orange in May and June.

The Mock-orange is extremely hardy, and will thrive in almost any soil or situation. It is a native of the South of Europe:—the dwarf variety, of Carolina.

This species may be increased by cuttings, planted early in October. They must always be kept tolerably moist. The other kinds may be increased in the same way, but must be sheltered in the winter season.

Mason speaks of the Syringa in his English Garden;

but it is doubtful whether he alludes to the species called the Mock-orange:

> " The sweet syringa, yielding but in scent
> To the rich orange; or the woodbine wild,
> That loves to hang on barren boughs remote
> Her wreaths of flowery perfume."

Some readers have supposed Mason's meaning to have been, yielding *in scent but* to the rich orange, &c.—and even then, he can scarcely be thought to do justice to the Syringa, if he means this species.

The lilac tree is called Syringa by the botanists, but has no connexion with this Syringa. Cowper mentions both, and the lines are so much to the purpose here, that although a part of the passage has been quoted in another part of the work, we must be allowed to repeat it:

> ———————" Laburnum, rich
> In streaming gold; syringa, ivory pure;
> The scentless, and the scented rose; this red,
> And of an humbler growth, the other tall,
> And throwing up into the darkest gloom
> Of neighbouring cypress, or more sable yew,
> Her silver globes, light as the foamy surf
> That the wind severs from the broken wave;
> The lilac, various in array, now white,
> Now sanguine, and her beauteous head now set
> With purple spikes pyramidal, as if
> Studious of ornament, yet unresolved
> Which hue she most approved, she chose them all."

It is very singular that Cowper makes no mention of the fragrance of the Syringa. Nothing can be more just than his description as far as it goes; but its exquisite beauty deserved more lingering over: had it been less beautiful, probably, more might have been said of its sweetness. Few flowers are more worthy of a poet's pen.

TAGETES.

CORYMBIFERÆ. SYNGENESIA POLYGAMIA SUPERFLUA.

This genus comprises the African and French Marygolds of the gardeners.—*French,* oeillet d' Inde [Indian pink]; rose d' Inde [Indian rose]; fleur de Rome [flower of Rome]; l'Africaine [the African].—*Italian,* tagete; garofano Messicano [Mexican pink]; garofano Africano [African pink]; garofano Turchesco; garofano d'India; fior di morto [death-flower].

THE African and French Marygolds belong to the genus Tagetes, so named from Tages, the grandson of Jupiter, and son of Genius, who first taught the Etruscans the art of divination.

The colour of the French Marygold varies from a bright yellow to a deep red orange-colour, and is often variegated with both. The scent is disagreeable: it flowers from the beginning of July until the frost checks it.

Of the African Marygold there are several varieties, also differing in colour. One is sweet-scented, and Parkinson observes "that it has the smell of a honeycomb, and is not of that poisonful scent of the former kinds."

These plants should be raised in a hot-bed; but are well worth purchasing for their rich and beautiful colours. They may be brought into the open air early in May, and will continue a long time in beauty; but both are annuals. They may be treated as the Common Garden Marygold.

TARCHONANTHUS.

AFRICAN FLEABANE.

CORYMBIFERÆ. SYNGENESIA POLYGAMIA ÆQUALIS.

ALL the species of this plant are from the Cape of Good Hope. The shrubby kind is the handsomest. The flowers are of a dull purple, and make little show: they begin to blow in the autumn, and continue to the end of the winter. The leaves, which are on all the year, are downy, and white underneath: they smell like bruised Rosemary-leaves.

This plant may be increased by cuttings planted in May: they should be kept within doors till the end of June; and all the plants, young and old, should be in the house from October till May. They should be shifted into fresh earth every year, and when requisite into larger pots.

All the species may be treated in the same manner: they are very thirsty plants, and must be allowed plenty of water.

TOBACCO-PLANT.

NICOTIANA.

SOLANEÆ. PENTANDRIA MONOGYNIA.

This genus is named from Jean Nicot of Nismes, agent from the King of France to Portugal, who procured the seeds from a Dutchman, and sent them to France. Tobacco, from the island Tobago. The French have many names for it; as, le tabac; nicotiane; petum, from its first introducer; herbe du grand prieur; herbe à la reine [the queen's herb]; Medicée [from the queen's family name]; buglosse antarctique; panacée antarctique [southern all-heal]; herbe sainte; herbe sacrée [holy herb]; herbe propre à tous maux [herb fit for all diseases]; jusquiame de Perou [Peruvian henbane]; herbe de Tournabon; herbe de St. Croix; herbe de l'ambassadeur.—*Italian*, tabacco; ternabona.

The Tobacco-plant is admitted into flower-gardens chiefly for its symmetrical growth, and luxuriant foliage; and some of the kinds are very handsome. The Broad-leaved Virginian or Sweet-scented Tobacco grows to the height of three or four feet; the leaves are ten inches long, and three and a half broad, and the blossoms of a deep purple.

This plant is usually raised in a hot-bed; but if sown in

March, and kept within doors for a month or two, it will grow very well Early in May it may be gradually inured to the open air; and, at the end of the month, may be removed carefully, with the ball of earth attached to it, into a large pot. It will require frequent watering; in small quantities while young, but when grown pretty strong should have it plentifully as well as often. The flowers will appear in July, and continue till the frost stops them.

Tobacco is cultivated in the open fields in many parts of the continent; and might, doubtless, be grown to advantage in England, if it were not prohibited by act of parliament, under a heavy penalty, and the charges of pulling it up, which may be done by any justice of the peace. This prohibition, which was made for the encouragement of our American colonies, still continues in force, though the colonies are lost. Small attempts at planting Tobacco have been made from time to time, which promised success.

A plantation in the seventeenth century being found to thrive, Cromwell, probably at the desire of the Americans, is said to have sent a troop of horse to trample it down*.

The smoking of Tobacco is said to have been first introduced into England by Sir Walter Raleigh. In the house where he lived, at Islington, are his arms, with a Tobacco-plant on the top of the shield. Tobacco has been highly panegyrised by the poets: one now living indeed goes great lengths in its praise:

> "For thy sake, tobacco, I
> Would do any thing but die,
> And but seek to extend my days
> Long enough to sing thy praise."
>
> C. Lamb.

Spenser bestows on it the epithet divine: Belphœbe finds the Squire Timias wounded:

* See Miller's Gardener's Dictionary.

"Into the woods thenceforth in haste she went,
To seek for herbs that mote him remedy;
For she of herbs had great intendiment,
Taught of the nymph, which from her infancy
Her nursed had in true nobility:
There, whether it divine tobacco were,
Or Panachæa, or Polygony,
She found, and brought it to her patient dear,
Who all this while lay bleeding out his heart-blood near."

Some have been as warm in the censure of Tobacco as others have been in its praise: Cowper calls it a "pernicious weed," and is very severe upon it. Our Scotch king, James I., is well known to have entertained a great aversion to the use of this plant, and even proceeded so far as to write a book against it, under the title of *A Counterblast to Tobacco;* in which the royal author informed his subjects that smoking, or to use the language of the day, taking tobacco, "is a custome loathsome to the eye, hatefull to the nose, harmefull to the braine, dangerous to the lungs; and in the blacke stinking fume thereof, nearest resembling the horrible Stigian smoke of the pit that is bottomlesse." Many temporary ebullitions of spleen of this monarch against Tobacco are on record: among others, his declaration that if he were to "invite the devil to dinner, he should have three dishes; a pig; a pole of ling and mustard; and a pipe of Tobacco for digesture." It must be owned, that on its first introduction, our ancestors carried its use to an enormous excess, smoking even in the churches, which made Pope Urban VIII., in 1624, publish a decree of excommunication against those who used such an unseemly practice; and Innocent XII., in 1690, solemnly excommunicated all those who should take snuff or tobacco in St. Peter's church at Rome.

TUBEROSE.

POLIANTHES.

NARCISSEÆ. HEXANDRIA MONOGYNIA.

Polianthes is from the Greek, and signifies City-flower.—*French,* la tubéreuse; jacinthe des Indes [Indian hyacinth].—*Italian,* tuberoso; tubero Indiano [Indian bulb].

The Tuberose grows naturally in India, whence it was first brought to Europe. In the warmer parts of the European continent it thrives as well as in its native soil. In Italy, Sicily, and Spain, the roots thrive and propagate without care where they are once planted. The Genoese cultivate it, and send the roots annually to England, Holland, and Germany, where the climate is less congenial to it.

This plant has long been cultivated in English gardens for its extraordinary beauty and fragrance.

There are several varieties; one with double flowers, "which was obtained from the seed by Monsieur Le Cour, of Leyden in Holland, who for many years was so tenacious of the roots, even after he had propagated them in such plenty as to have more than he could plant, that he caused them to be cut in pieces, to have the vanity of boasting that he was the only person in Europe who possessed this flower; but of late years the roots have been spread into many parts*."

Those roots are the best which are the largest and plumpest, provided they are sound and firm, and the fewer offsets they have, the stronger they will flower. The under part of the roots should be particularly examined, because it is there that they first decay. Before the roots are

* See Miller.

planted, the offsets should be taken off, or they will draw a great deal of nourishment from the old root. They may be planted in April and May, but should be kept indoors, admitting fresh air in mild weather: most persons raise these flowers in a hot-bed, but the temperature of an inhabited room will generally bring them forward. They should be supported by sticks as the flower-stems advance in height, and should have little or no water till they begin to shoot; when in flower, they require plenty.

Flowers raised in this manner will blow about September and October, adorning and perfuming the apartment they are placed in, in a very agreeable manner. When the roots are strong, they will often produce ten or twelve flowers, and the stem will rise three or four feet high. As the flowers come out in spikes, opening successively from the bottom to the top, they will, of course, continue longer in beauty in proportion to the number they produce. They may be placed in a balcony in summer weather, if desired; but the double-flowered variety must remain in the room: if these are placed at a little distance from a closed window on which the sun shines (yet the room being properly ventilated), they will open more fair than when too much exposed.

The Malayans style the Tuberose the Mistress of the Night:

"The tuberose, with her silvery light,
That in the gardens of Malay
Is called the Mistress of the night,
So like a bride, scented and bright,
She comes out when the sun's away."

LALLA ROOKH.

We are to remember here that the poet is speaking of the lady's habits when in her native country; in our colder climate she must wait for the sunshine.

When worn in the hair by a Malayan lady, it informs her lover that his suit is pleasing to her.

TULIP.

TULIPA.

LILIACEÆ. HEXANDRIA MONOGYNIA.

Tulipa, from the resemblance of the flower to the eastern head-dress, called Tulipan, or Turban. Gerarde calls it Turk's-cap, or Dalmatian-cap, a name more commonly given to the Martagon-lily.—*French*, la tulipe.—*Italian*, tulipano.

Tulips are supposed to have been introduced into England about the year 1580. The kind commonly called the Garden Tulip has many varieties, which increase in number every year. In 1629 Parkinson enumerates 140 varieties: "But to tell you of all the sorts," says he, "which are the pride of delight, they are so many, and as I may say almost infinite, doth pass my ability, and, as I believe, the skill of any other. There is such a wonderful variety and mixture of colours in them, that it is almost impossible for the wit of man to decipher them thoroughly, and to give names that may be true, and several distinctions to every flower. Threescore several sorts of colours, simple and mixed, I can reckon up that I have, and of especial note; and yet I doubt not, but for every one of them there are ten others differing from them. But besides this glory of variety in colours that these flowers have, they carry so stately and delightful a form, and do abide so long in their bravery, that there is no lady or gentlewoman of any worth that is not caught with this delight." One of the earliest blowing varieties is the Duke Nanthol, which is in great estimation, as is also the Claremond; but it would be endless to attempt enumeration: all are esteemed. The best soil to plant Tulip roots in is a sandy earth, with the turf rotted amongst it: some add a fourth of sea-sand. The roots must be planted three or four inches deep, according to their size. The early-blowers should be planted in September, in a pot about ten or eleven inches deep; they should remain in-doors till April or May. If the weather be very scorching when they are in flower, they must be shaded in the heat of the day. When the flowers have decayed, and the seed-vessel begins to swell, it must be broken off; for if they are permitted to seed, the roots will be weakened thereby.

When the leaves of the early-blowers have decayed, which will be while the late blowers are yet in flower, the roots must be taken up, cleaned, spread in the shade to dry, and put away in a dry and secure place till they are wanted to plant again. The offsets from the roots, until they are big enough to flower, may be planted several together; but should be taken up when the leaves decay, the same as the old roots. These will flower early in the spring: the later blowers will flower a month or two later, in May and June; and these last may be planted in October or November. The roots should be taken up, and replanted every year, as directed above. In mild weather Tulips may stand abroad, and may be allowed to receive a soft shower, but must be screened from heavy rains; they will require little or no water, and, while in flower, must be sheltered from rains. The Garden Tulip is a native of the Levant; Linnæus says, of Cappadocia. It is very common in Syria; and is supposed, by some persons, to be the lily of the field alluded to by Jesus Christ. In Persia, where it grows in great abundance, it is considered as the emblem of perfect lovers. " When a young man presents one to his mistress," says Chardin, " he gives her to understand, by the general colour of the flower, that he is on fire with her beauty; and by the black base of it, that his heart is burnt to a coal." Chardin saw it on the northern confines of Arabia. Conrad Gesner first made the eastern Tulip known by a description and figure. Balbinus asserts that Busbequius brought the first Tulip-roots to Prague, whence they were spread all over Germany. Busbequius himself says, in a letter written in 1754, that this flower was then new to him. We know that he collected natural curiosities, and brought many from the Levant. He relates that he paid very dear to the Turks for Tulips; but he nowhere affirms that he was the first who brought them from the East. In 1565, there were Tulips in the garden of Mr. Fugger, from whom Gesner wished to procure some. The first Tulips planted in England were sent from Vienna about the end of the sixteenth century *.

These flowers, of no further utility than to ornament gardens, and whose duration is short and very precarious, became, in the middle of the seventeenth century, the object of a trade for which there is no parallel, and their price rose beyond the value of the most precious metals. Many authors have given an account of this trade, some of whom have misrepresented it. Menage called it the Tulipomania, at which people laugh because they believe that the beauty and rarity of the flowers induced florists to give such extravagant prices. But this Tulip-trade was a mere gambling commerce, and the Tulips themselves were only nominally its objects: many bargains being daily made, and the roots neither given nor received. A long and curious account of this trade is to be found in the first volume of Beckmann's History of Inventions.

Persons fond of flowers, however, particularly in Holland, have paid very high prices for Tulips, as the catalogues of flowers show. In the year 1769 the dearest

* See Beckmann's History of Inventions, vol. i.

kinds in England were the Don Quevedo and the Valentinier: the former was sold at two guineas; the latter at two and a half.

"This," says Beckmann, "may be called the lesser Tulipomania, which has given occasion to some laughable circumstances. When John Balthasar Schuppe was in Holland, a merchant gave a herring to a sailor who had brought him some goods. The sailor seeing some valuable Tulip-roots lying about, which he considered as of little consequence, thinking them to be onions, took some of them unperceived, and ate them with his herring. Through this mistake the sailor's breakfast cost the merchant a much greater sum than if he had treated the Prince of Orange."

"Another laughable anecdote is told of an Englishman, who, being in a Dutchman's garden, pulled a couple of Tulips, on which he wished to make some botanical observations, and put them into his pocket; but he was apprehended as a thief, and obliged to pay a considerable sum before he could obtain his liberty."

In proportion as Tulips blow later in the year, their stems are longer, and consequently the more they require support: their bending to the wind, and their resemblance to the turbans from which they are named, are alluded to by Mr. Moore in the following lines:

"What triumph crowds the rich Divan to-day
With turban'd heads, of every hue and race,
Bowing before that veiled and awful face,
Like tulip-beds, of different shape and dyes,
Bending beneath th' invisible west wind's sighs!"

LALLA ROOKH.

A Turkish poet, in an ode translated by Sir W. Jones, compares "Roses and Tulips to the cheeks of beautiful maids, in whose ears the pearls hang like drops of dew."

"Then comes the tulip-race, where beauty plays
Her idle freaks: from family diffused
To family, as flies the father dust,
The varied colours run; and while they break
On the charmed eye, th' exulting florist marks
With secret pride the wonders of his hand."

THOMSON.

VALERIAN.

VALERIANA.

VALERIANEÆ. TRIANDRIA MONOGYNIA.

The derivation of this name is uncertain.—*French,* la valériane.—*Italian,* valeriana.

The Valerians vary in size from three or four feet to as many inches; their flowers are commonly red or white, but there are a few species with blue, and with yellow flowers.

The seeds may be sown of the annual kinds, and the roots parted of the perennial, in spring or autumn. Some of them, as the Red and the Alpine Valerians, thrive best on rocks, old walls, or buildings; the seed being scattered in the joints and chinks.

The Pyrenean species likes shade and a moist soil: the Garden Valerian likes moisture too, and plenty of room, as it spreads fast.

All the kinds must be kept moderately moist. Some give the Alpine kinds a poor stony soil covered with moss, in imitation of their natural place of growth, on mossy rocks, where the snow lies six or seven months in the year.

VERVAIN.

VERBENA.

VERBENACEÆ. DIDYNAMIA ANGIOSPERMA.

The derivation of Verbena is uncertain: it originally signified any herb used to decorate altars. The present plant is also named Juno's-tears, Columbine, and Pigeon's-grass.—*French,* verveine.—*Italian,* verbena.

The Verbenas are generally natives of warm countries, and require much care and tenderness; most of them may be preserved, however, without a stove, when once raised. The Cut-leaved Rose Vervain is an annual or biennial plant, in some estimation for its brilliant colours. It flowers in June and July.

The most popular kind is the Three-leaved, of which the scent seems to partake of the Lemon and the Almond. The leaves are delicate and elegant; the flowers pale purple. This delightful little shrub is a native of South America: it may stand abroad in the summer, but should be housed again about Michaelmas. It may be increased by cuttings planted in any of the summer months.

The Common Vervain—in French, *vervene verveine; herbe sacrée* [sacred herb]: in Italian, *verbena; erba colombina* [dove-wort]—is a native of Europe, Barbary, China, Cochinchina, and Japan. With us it grows by road-sides, and in dry sunny pastures. Mr. Miller remarks, "that although Vervain is very common, yet it is never found above a quarter of a mile from a house," whence it has been named by some, Simpler's-joy. The fact, however, is not allowed; and Dr. Withering found it in plenty at the foot of St. Vincent's rocks. It begins to flower in July, and continues to the end of autumn.

Vervain was held sacred among the ancients, and was employed in sacrifices, incantations, &c.: it is one of the plants termed by the Greeks Sacred Herb. It was suspended round the neck as an amulet, thought good against venomous bites, and recommended as a sovereign medicine for various diseases. In Britain it has fallen into disuse, in spite of a pamphlet written expressly to recommend it, directing the root to be tied with a yard of white satin riband round the neck, and to remain there till the patient recovered.

Drayton, in the Muse's Elysium, calls it the "Holy Vervain;" and in the same poem speaks of it as worn by heralds:

> " A wreath of vervain heralds wear,
> Amongst our garlands named,
> Being sent that dreadful news to bear,
> Offensive war proclaimed."

> " Black melancholy rusts, that fed despair
> Through wounds long rage, with sprinkled vervain cleared;
> Strewed leaves of willow to refresh the air,
> And with rich fumes his sullen senses cheered."
>
> DAVENANT'S GONDIBERT.

VIOLET.

VIOLA.

VIOLEÆ. PENTANDRIA MONOGYNIA.

French, violette.—*Italian*, viola.

THE species of Violets are very numerous; the Tricoloured or Pansy-violet has been noticed under its more familiar name of Heart's-ease. At the head of the other Violets ranks the Viola Odorata, or Sweet-violet—in French, *violette de Mars* [March violet]; *violier commun:* and in Italian, *viola Marzia; viola mammola*—which is a native of every part of Europe, in woods, bushes, and hedges, flowering in March and April. In 1804, Mr.

Martyn gathered a handful of them from one root at the end of November. The flower varies in colour, though most commonly a deep purple: it is sometimes of a paler purple, sometimes a red-purple, flesh-coloured, or quite white; but it is always delightfully fragrant.

A syrup is prepared from this Violet, used in chemistry to detect an acid or an alkali, the former changing the blue colour to a red, and the latter to a green. For this purpose, the flowers are cultivated in large quantities at Stratford upon Avon. This Violet is very common in Japan, where it flowers from January to April. There are both double and single-flowered varieties: the white are generally the largest flowers; some maintain them to be the most fragrant: they blow later than the purple.

The Violet is scarcely less a favourite with the poets than the Rose itself; but whether lovers will acquiesce in the assertion made by Mr. Barry Cornwall, they must decide.

"There was a mark on Laïs' swan-like breast,
(A purple flower with its leaf of green,)
Like that the Italian saw when on the rest
He stole of the unconscious Imogene,
And bore away the dark fallacious test
Of what was not, although it might have been,
And much perplexed Leonatus Posthumus:
In truth, it might have puzzled one of us.

"The king told Gyges of the purple flower;
(It chanced to be the flower the boy liked most;)
It has a scent as though love, for its dower,
Had on it all his odorous arrows tost;
For though the rose has more perfuming power,
The violet, (haply 'cause 'tis almost lost,
And takes us so much trouble to discover)
Stands first with most, but always with a lover."

———————"such odours as the rose
Wastes on the summer air, or such as rise

From beds of hyacinths, or from jasmine flowers,
Or when the blue-eyed violet weeps upon
Some sloping bank remote, while the young sun
(Creeping within her sheltering bower of leaves)
Dries up her tears."

BARRY CORNWALL.

The Violet seems a favourite with this author: he introduces it continually. In his last poem, the Flood of Thessaly, he mentions it several times:

"And violets, whose looks are like the skies."

"Jasmine and musk, daisies and hyacinth',
And violets, a blue profusion, sprang
Haunting the air."

The Violet is continually applauded for its modesty and timidity:

————"steals timidly away,
Shrinking as violets do in summer's ray."

LALLA ROOKH.

Mr. Keats delights in describing a little woodland nook, and Violets constantly breathe their sweet perfume in it. —(See HAWTHORN.)

————"where to pry aloof,
Atween the pillars of the sylvan roof,
Would be to find where violet beds were nestling,
And where the bee with cowslip-bells was wrestling."

"Gay villagers, upon a morn of May,
When they have tired their gentle limbs with play,
And formed a snowy circle on the grass,
And placed in midst of all that lovely lass
Who chosen is their queen;—with her fine head
Crowned with flowers, purple, white, and red;
For there the lily and the musk-rose sighing,
Are emblems true of hapless lovers dying:
Between her breasts, that never yet felt trouble,
A bunch of violets, full-blown, and double,
Serenely sleep."

Ebn Abrumi, an Arabian poet, compares blue eyes weeping, to Violets bathed in dew *.

How beautiful is the following passage in the Winter's Tale!

> ————"violets, dim,
> But sweeter than the lids of Juno's eyes,
> Or Cytherea's breath."

In Cymbeline, Belisarius, speaking of the two young princes, says,

> ————"They are as gentle
> As zephyrs, blowing below the violet,
> Not wagging his sweet head:"

In Twelfth Night again, the poet has some exquisite lines upon this flower, where the duke, listening to plaintive music, desires

> "That strain again; it had a dying fall:
> O, it came o'er my ear like the sweet south,
> That breathes upon a bank of violets,
> Stealing, and giving odour."

We are told, in the notes to Mr. Steevens' Edition of Shakspeare, that the Violet is an emblem of faithfulness: to corroborate which, he gives some lines from a sonnet, published in a collection printed in the year 1584:

> "Violet is for faithfulnesse
> Which in me shall abide;
> Hoping likewise that from your heart
> You will not let it slide."

Burns speaks of the hyacinth as an emblem of fidelity: its virtue lies, it seems, in the colour, and may be extended to all flowers of *true blue*. The insertion of the song will be readily forgiven me.

> "O luve will venture in, where it daur na weel be seen,
> O luve will venture in, where wisdom ance has been;
> But I will down yon river rove, amang the wood sae green,
> And a' to pu' a posie to my ain dear May.

* See Carlisle's Specimens of Arabian Poetry, 75.

"The primrose I will pu', the firstling of the year,
And I will pu' the pink, the emblem of my dear;
For she's the pink o' womankind, and blooms without a peer;
And a' to be a posie to my ain dear May.

"I'll pu' the budding rose, when Phœbus peeps in view,
For it's like a balmy kiss o' her sweet bonnie mou;
The hyacinth's for constancy, wi' its unchanging blue;
And a' to be a posie to my ain dear May.

"The lily it is pure, and the lily it is fair,
And in her lovely bosom, I'll place the lily there;
The daisy's for simplicity and unaffected air;
And a' to be a posie to my ain dear May.

"The hawthorn I will pu' wi' its locks o' siller gray,
Where, like an aged man, it stands at break o' day;
But the songster's nest within the bush I winna take away;
And a' to be a posie to my ain dear May.

"The woodbine I will pu', when the ev'ning star is near,
And the diamond drops o' dew shall be her een sae clear;
The violet's for modesty, which weel she fa's to wear,
And a' to be a posie to my ain dear May.

"I'll tie the posie round wi' the silken band o' luve,
And I'll place it in her breast, and I'll swear by all abuve,
That to my latest draught of life, the band shall ne'er remove;
And this shall be a posie to my ain dear May."

Another of our rustic poets, Clare, has a poem addressed to the Violet, in the second volume of his Village Minstrel, &c.; in the first volume, in a poem entitled Holywell, he speaks of it as one of the first signs of Spring:

"And just to say the Spring was come
The violet left her woodland home,
And, hermit-like, from storms and wind
Sought the best shelter it could find,
'Neath long grass banks."

Mr. Moore, in his notes to Lalla Rookh, quotes some passages to inform us that the Sweet Violet is one of the plants most esteemed in the East, particularly for its use in sherbet; which they make with violet sugar. "The sherbet they most esteem, and which is drank by the grand signor himself, is made of violets and sugar.—TAVERNIER."

Mr. H. Smith, in his Amarynthus, speaks of this flower as being of short duration.

> —" the trembling violet, which eyes
> The sun but once, and unrepining dies."

The North American Violets are mostly void of scent, with the exception of the Dog's-violet, with which we are also familiar in our own hedges, as a successor to the Sweet-violet. With this exception too, the North American Violets best succeed in loam and bog earth, and should be housed in the winter.

VIPER'S BUGLOSS.

ECHIUM.

BORRAGINEÆ. PENTANDRIA MONOGYNIA.

This plant has been supposed to cure the bite of the viper: it is also called cat's tail.—*French,* la viperine; l'herbe aux viperes [viper's wort].—*Italian,* viperina.

THE Cretan species is the handsomest of the genus: its flowers are of a red-purple: the plant produces them but once. This kind is a native of the Levant: its stalks are trailing, and about a foot in length. The top of a wall is the best place to sow it; if in a pot, it must be in a gravelly soil: it should be sown about the middle of October, and in hard frost covered with a little sawdust, straw, or oak-leaves. It will flower in July and August; and if on a wall, will scatter its own seeds, and so maintain its own continuance.

The other species must be housed in the winter: they do not produce their flowers till the second year after sowing. They must be sparingly watered, in winter particularly; the stems being succulent.

VIRGINIA COWSLIP.

DODECATHEON MEADIA.

PRIMULACEÆ. PENTANDRIA MONOGYNIA.

French, gyroselle de Virginie.

This is a perennial plant, with purple flowers, or inclining to the colour of the peach blossom. It is very ornamental when in flower, which is in April and May.

This plant is more impatient of heat than of cold: it will endure our most severe winters; but two or three days' exposure to a hot sun will entirely destroy the young plants. It may be increased by offsets from the roots, which should be taken off, and transplanted in August, after the leaves and stalks have decayed, that they may have time to gain strength before the frost comes on.

WALLFLOWER.

CHEIRANTHUS.

CRUCIFERÆ. TETRADYNAMIA SILIQUOSA.

French, giroflier jaune; violier jaune [both signifying yellow stock]; le baton d'or [gold stick]; la ravanelle; le rameau d'or [golden branch]; le garranier jaune.—*Italian,* viola [stock]; viola gialla [yellow stock]; cheiri.

The Wallflowers are, in fact, Stocks; since they not only belong to the same genus, but are properly named Wallflowers, or Stocks: but some of the species having been distinguished by custom as Wallflowers, entirely dropping the other name, they are here placed under that head.

Of all the species so named, the common Wallflowers are by far the finest; their colours are extremely rich, and, as the artists express it, warm; and their fragrance very delicious: they are apt to have a ragged appearance, looking sometimes at a little distance like a number of beautiful petals hung accidentally together; but when their form is preserved, they are in every respect elegant. There are single and double varieties; red and yellow of all shades, and pure white. The flower is too well known to make further description necessary.

"The common Wallflower," says Mr. Martyn, "is a native of Switzerland, France, Spain, &c.; and is common on old walls and buildings in many parts of England. It is one of the few flowers which have been cultivated for their fragrancy time immemorial in our gardens."

Some prefer the Alpine Wallflower for appearance, the flowers being usually larger, and closer together; but they have not so fine a scent as the common Wallflower.

If raised from seeds, they should be sown in April, two or three seeds in a middle-sized pot; and in a poor rubbishy soil. If the soil be poor, they will bear the winter abroad; and will flower the following June.

The double varieties are increased by slips, about three inches long, planted in the spring; they should be slipped off with a sharp knife, and one-third inserted in the earth, the leaves being stript from the lower half.

Early in September the seedlings should be transplanted into separate pots. Some persons sow them where they are to remain; but transplanting is generally supposed rather to benefit than to injure them.

"Fair handed Spring unbosoms every grace;
Throws out the snow-drop, and the crocus first,

The daisy, primrose, violet darkly blue,
And polyanthus of unnumbered dyes;
The yellow wallflower, stained with iron brown;
And lavish stock, that scents the garden round."

THOMSON'S SPRING.

WATER LILY.

NYMPHÆA.

NYMPHÆACEÆ. POLYANDRIA MONOGYNIA.

Called also water-rose, water-can.—*French,* lis des étangs, [pond lily], volet, plateau, jaunet d'eau [water yellow-flower].—*Italian,* nenufaro, ninfea, blefera.

THE Water-lilies cannot be grown but in a cistern of water, which should be lined with lead; and such plants are only adapted for persons having even a superfluity of garden ground. For such persons, they are very desirable, for they are delicate and elegant plants:

"Those virgin lilies, all the night
Bathing their beauties in the lake,
That they may rise more fresh and bright
When their beloved sun's awake."

MOORE'S LALLA ROOKH.

"And now the sharp keel of his little boat
Comes up with a ripple, and with easy float,
And glides into a bed of water lilies:
Broad-leaved are they, and their white canopies
Are upward turn'd to catch the heaven's dew.
Near to a little island's point they grew;
Where Calidore might have the goodliest view
Of this sweet spot of earth."

KEATS.

The Japanese set a high value upon the Water-lily, because of its purity, not being sullied by contact with the muddy water, in which it often grows*. This Water-lily is said to be the ancient herb Lotus; which, with the Crocus and the Hyacinth, formed the couch of Jupiter and Juno; and yet Achilles was so profane as to feed his horses with it. It is not to be understood as the Lotus which gave name to the Lotophagi. That was a tree (for the ancients had both a herb and a tree so named) now called the Rhamnus Lotus.

Southey mentions the herb Lotus in his Curse of Kehama:

> "The large-leaved lotus on the waters flowering."
>
> Vol. i. page 86.

In Japan the Water-lily (there called Tratte), being, for the reason before mentioned, considered as an emblem of purity, is, with the flowers of the Motherwort, borne in procession before the body in their funeral ceremonies: these are carried in pots: artificial Water-lilies of white paper are also borne on poles†.

Moore, in his notes to Lalla Rookh, observes, that in some parts of Asia the women wear looking-glasses on their thumbs: "Hence," says he "(and from the lotus being considered the emblem of beauty) is the meaning of the following mute intercourse of two lovers before their parents:

> "He with salute of deference due
> A lotus to his forehead prest;
> She raised her mirror to his view,
> Then turned it inward to her breast."

In another part of the same poem, Moore compares the eyes of Love to the blue Water-lily:

> "And his floating eyes—oh! they resemble
> Blue water-lilies, when the breeze
> Is making the stream around them tremble."

* See Titsingh's Illustrations of Japan. † Ibid.

This blue species is a native of Cashmere and Persia. Mrs. Graham, in her residence in India, speaks of a beautiful red lotus, of which she saw multitudes; she describes them as much larger than the White Water-lily, and the loveliest of the nymphæas she had ever seen.

WINGED-PEA.

LOTUS TETRAGONOLOBUS.

LEGUMINOSÆ. DIADELPHIA DECANDRIA.

French, le lotier rouge [red lotus].

The Winged-Pea (a species of the Lotus, or Bird's-foot Trefoil) is a native of Sicily: its colours are scarlet and purple. It is sometimes called the Scarlet Pea, the Crimson-velvet Pea, the Square-codded Pea, &c. This also flowers in June and July. This may be treated in the same manner as the Sweet Pea, which is the chief of the Garden Peas.

WINTER-CHERRY.

PHYSALIS.

SOLANEÆ. PENTANDRIA MONOGYNIA.

This plant is also named Alkekengi.—*French,* coqueret; coquerelle; quoquerelle; herbe a cloques.—*Italian,* alchechengi; alcachingi; solano alicacabo.

The Winter-Cherry has not much beauty, except in the autumn, when it is in fruit. It is a native of the South of Europe, Germany, China, and Cochin-China. It may be increased by parting the roots after the stalks have decayed. This plant loves the shade, and the roots require confinement.

In Spain, Switzerland, and some parts of Germany, the country-people eat these cherries by handfuls: here they are only cultivated for their beauty.

A species of the Solanum, or Night-shade—*solanum pseudocapsicum*—is now more commonly known by the name of Winter-Cherry, and in France by those of *morelle cerisette, petit cerisier d'hiver, amome des jardiniers;* of which, also, the fruit in appearance resembles the Cherry. It requires shelter from severe frost, and therefore should be housed, but not kept too warm. It should every year, in the month of April, be taken out of the pot; all the decayed and matted roots on the outside should be cut off, and it should be filled up with fresh rich earth. This treatment will greatly improve the flowers and fruit.

The earth must be kept tolerably moist for both these plants.

The latter plant exhibits its blossoms and fruits both at the same time, as the latter remain on the shrub all the winter. The fruit is supposed to be poisonous, but it has been tried upon a dog without producing any ill effect.

> " The amomum there with intermingling flowers
> And cherries hangs her twigs."
>
> COWPER'S TASK.

XERANTHEMUM.

CORYMBIFERÆ. SYNGENESIA POLYGAMIA SUPERFLUA.

This name is Greek, and signifies a dry-flower.—*French,* l'immortelle.

THE Annual Xeranthemums should be sown in the

autumn, singly, in a pot of five or six inches diameter, filled with light earth. If in a warm situation, as near a wall facing the south or south-east, they will bear an ordinary winter abroad: in June they will begin to flower, and in July are fit for gathering. There are single and double varieties; and, contrary to the habit of most double flowers, these may generally be continued from seed. The flowers are commonly white or purple.

The other kinds are chiefly shrubby, and propagated by cuttings, planted in any of the summer months in a pot of light earth. These kinds must be sheltered from frost; but, if not allowed fresh air in mild weather, will grow up weakly, and often bear no flowers. They should be placed near a window, open in mild weather. In dry summer weather the Xeranthemums will require frequent watering, but must be sparingly watered in winter.

These flowers, if gathered in their beauty, will preserve it many years, and make a showy figure with other dried flowers, as Amaranths, Honesty, Gnaphaliums, &c. in the winter. They also make pretty ornaments for a lady's hair: their colours are white, purple, yellow, or red.

The Xeranthemum has of late been highly improved by culture, and many persons are very curious in them.

YUCCA.

LILIACEÆ. HEXANDRIA MONOGYNIA.

Frequently called Adam's-needle.—*French,* yuca; youc.

The Superb Yucca is from North America; it was first cultivated in Europe by Gerarde, to whom it was brought from the West Indies by the servant of an apothecary.

Gerarde kept the plant till his death; Parkinson had it from the widow, and with *him* it perished.

The Yucca is nearly allied to the Aloe, and, like that, blows very rarely: the flowers of this species are bell-shaped, and hang downward; they are white within, but on the outside each petal has a stripe of purple; they appear in August and September.

The Aloe-leaved Yucca is a native of South America: it produces a greater abundance of flowers than the former kind; white on the inside, purple without.

The Drooping-leaved Yucca has white flowers, but its scent is not agreeable.

The Thready Yucca, so called from long threads which hang from the sides of the leaves, is a native of North America. The flower-stem of this plant grows to the height of five or six feet, and nearly the whole of it is covered with large white flowers, sitting close. But, like the other kinds, this plant flowers but seldom.

It is said that, about the middle of the seventeenth century, Mr. Walker possessed abundance of these plants in his suburban garden in the village of St. James. "But," says Mr. Morison, who mentions the circumstance, "I never saw it flower there."

These plants are to be treated as hardy Aloes.

ZINNIA.

CORYMBIFERÆ. SYNGENESIA POLYGAMIA SUPERFLUA.

So named by Linnæus, in honour of J. G. Zinn, pupil of Haller, and professor of botany at Gottingen after him.

The Zinnias are annual plants, bearing handsome flowers: they are usually raised in a hot-bed; but a warm inhabited room will generally bring them forward as well.

The Yellow Zinnia is the most tender; and this will produce but few flowers, unless it is stinted in its growth while young, by confining the roots in a small pot. It is a Peruvian.

The Red Zinnia is a native of North America; when in full beauty the flower is of a red-purple, powdered with gold, like one of the species of the Amaryllis; it afterwards becomes more dull, red and yellow, and green underneath. There is a variety with yellow flowers.

The Whorl-flowered and the Purple Zinnias are from Mexico: the first has double red flowers; the latter handsome flowers, first red, but changing to a deep violet. These, and the Slender-flowered, which is a native of South America, with orange-coloured flowers, may be gradually accustomed to the open air about the end of May: in July they will begin to blow, and continue in bloom till the approach of frost.

The seeds should be sown in March, singly;—or several together, and towards the end of May transplanted into separate pots, of ten or twelve inches diameter. The earth must be kept moderately moist; but water must be given in sparing portions in winter, rather in sips than draughts.

ZYGOPHYLLUM.

ZYGOPHYLLEÆ. DECANDRIA MONOGYNIA.

Often called Bean-caper.—*French,* fabagelle.

THE Scarlet-flowered Bean-caper is a native of Africa and Siberia; the White, of Egypt; the Four-leaved, and Sessile-leaved, from the Cape, have yellow flowers. They are succulent plants, and must be sheltered from surrounding damps, and in the winter from the cold.

GENERAL OBSERVATIONS.

THE observations necessary to make here will be very few, and will only comprise such information as a person should have in memory, who attempts to rear plants in any way.

SOWING.

Where nothing is said to the contrary, it may be taken as a general rule to sow seeds in proportion to their size, from a quarter of an inch to an inch deep. The Convolvulus, and such sized seeds in general, may be sown a quarter of an inch deep; the Lupine, &c., half an inch; the Scarlet-bean, &c., an inch deep. A few pebbles should be put at the bottom of the pot, to drain off superfluous moisture: the soil will, of course, vary according to the plant: but whatever soil the plant may require, and many require a stony one, the earth should be light and free from stones above the seeds. After sowing, a little water should be given to settle the earth about the seeds. It is not advisable for individuals to save their seeds from home-reared plants: besides that they thrive better in a change of soil, they will often be stronger from plants in the open ground.

PLANTING.

To see if a plant wants fresh potting, turn it carefully out of the pot, with the earth attached to it, and examine the roots. If they are matted about the sides and bottom of the ball, the plant evidently requires fresh potting. Then carefully reduce the ball of earth to about a third of its original bulk; single out the matted roots, and trim away all that are mouldy and decayed. Probably the same pot may then be large enough; but, if it requires a larger, it should be about two inches broader for a middle-sized plant; three or four for a large plant. If the roots are not matted, but the pots are filled with the fibres, keep the ball entire, and carefully plant it in a larger pot. At the top of a large pot, an inch; of a small pot, half an inch, should be left for the reception of water, without danger of overflow. A little gravel should always be at the bottom.

A plant newly potted must never be exposed to a strong sun: it should be watered and placed in the shade immediately, and there remain till it is rooted; which may be known by its shooting above.

Plants are frequently destroyed by replanting, merely from the careless manner in which it is done. Where the roots spread, plenty of room should be left open, a little hillock made in the centre of the pot, and the plant being placed thereon, the roots should be distributed around it in a regular manner, observing that they are not twisted or turned up at the ends. The earth should be filled in, a little at a time, and the pot gently shaken, to settle the earth to the roots all the way down. When filled, it should be pressed down with the hand. It is very common to fill in the earth at once, and press it hard down; which not only wounds the tender fibres, but often leaves a hollow space towards the bottom of the roots, and deprives them of their proper nourishment. But the thing most necessary to be observed is, *that the roots be allowed their natural course.*

All plants should be kept clear of weeds, not for neatness alone, but because they exhaust the nutriment which should feed the plant.

WATER.

The best water for plants is undoubtedly rain-water; if this cannot be obtained river-water will do: pond-water is not so good; but the worst of all is hard spring water. In winter, and, for delicate plants, even in summer, water should be placed in the sun till it becomes tepid before it is used.

The water should never be allowed to remain in the pan under the pot; it tends to rot the roots. It may be well to observe that plants should be watered with a rose on the spout of the watering-pot; and the more finely it is perforated the better, so as to sprinkle the water lightly over the flowers and leaves, without bending them down with its weight.

> " E spesso irrigherai le lor radici,
> Prendendo un vaso di tenace creta
> Forato a guisa d'un minuto cribro,
> Che i Greci antichi nominar clepsidra,
> Per cui si versan fuor milla zampilli.
> Con esso imitar puoi la sottil pioggia,
> Ed irrorar tutte le asciutte erbette."
>
> LE API DEL RUCELLAI.

" And you should often water their roots; take a vessel of hardened clay perforated in the manner of a fine sieve, such as the ancient Greeks called a clepsidra; through which may be shed a thousand streams. With this you may imitate a light shower, and water all the dry herbs."

Many persons think it sufficient to water the roots, which is a great mistake; it materially contributes to their health and beauty to sprinkle the whole plant. Bathing the feet will neither cleanse the hands, nor freshen the bloom of the cheeks. The Auricula, however, is one of those ladies of fashion, who, fearful of injuring the delicacy of their complexions by the use of water, cleanse their faces with some elegant powder as a substitute for that rude element. There is a farina in the flower of the Auricula, which is usually esteemed its principal beauty: this is the case with the Polyanthus also, and some few others, of which the blossoms must not be watered.

Of such plants as are succulent, it is generally advised to water the leaves but seldom, lest a redundancy of moisture should rot them: the merely plucking a leaf will generally determine whether a plant be succulent or not; and a person may distinguish in a moment what blossoms are likely to be hurt by water, by observing whether there be any visible farina on them. The best way in watering all plants, is rather to cast the water at, than to pour it on them, as it falls more lightly. It will be observed that more water, as well as more shelter, must be necessary for potted plants than for plants in the open ground.

AIR AND LIGHT.

Flowers must not be denied the light, towards which they naturally turn; the want of it will injure their health as much as the want of water, air, or warmth.

They must also be allowed air: even those that will not bear the outer air must have the air of the room frequently freshened by ventilation, to preserve them in health. Care should be taken not to let plants stand in a draught; for, so situated, one strong gust of an easterly wind will often prove sufficient to destroy them.

In frosty weather the windows should be kept close, and at night the shutters. Those plants directed to be placed in an inhabited room will not usually require a fire, if in a room which has had fire in the day; but in sharp frost it will be well, instead of stirring out the fire, to leave a little, on retiring to rest, and place a guard before it for security.

INSECTS.

If any plants are infested with insects, which is often the case with Rose-trees, Heart's-ease, &c., they should be watered with tobacco-water, which quickly destroys them. Some say that, independently of the removal of the insects, it improves the verdure of the plant.

One pound of roll-tobacco will suffice for three pints of water, which should be poured on it nearly boiling, and stand a few hours before it is used.

BULBS.

The leaves of bulbous flowers should never be plucked before they decay, or the bulb will thereby be deprived of a large portion of its natural nourishment. When the flowers and leaves have decayed, the bulbs should be taken up, dried in the shade, all loose earth, fibres, &c. should be cleaned off them, and they should be put away in a dry place, and safe from mice, &c., until wanted. They should not touch each other, but either lie all on a flat surface, or be kept apart by some dry sand. Bulbs should have no fresh water after their leaves have began to decay.

Any person having too many bulbs of one kind, and too few of another, will find no difficulty in obtaining an exchange at the shops where they are sold.

Some persons put a piece of nitre, of the size of a pea, into the water in which bulbs are raised, renewing it whenever the water is changed, in order to make the colours brighter. It may not be known to every one who finds amusement in the cultivation of flowers, that pots are made purposely for bulbs, as wide at the bottom as at the top, and of the requisite depth. When they are grown in earth, it is advisable to procure these pots; for of the common kind they cannot be obtained sufficiently deep without being much wider than necessary.

THE END.

LONDON:
PRINTED BY THOMAS DAVISON, WHITEFRIARS.

Printed in the United States
By Bookmasters